Generic Data Structures and Algorithms in Go

An Applied Approach Using Concurrency, Genericity and Heuristics

Richard Wiener

Apress®

Generic Data Structures and Algorithms in Go

Richard Wiener
Colorado Springs, CO, USA

ISBN-13 (pbk): 978-1-4842-8190-1
https://doi.org/10.1007/978-1-4842-8191-8

ISBN-13 (electronic): 978-1-4842-8191-8

Managing Director, Apress Media LLC: Welmoed Spahr
Acquisitions Editor: Steve Anglin
Development Editor: Jim Markham
Coordinating Editor: Gryffin Winkler

Cover image designed by Freepik (www.freepik.com)

Distributed to the book trade worldwide by Springer Science+Business Media LLC, 1 New York Plaza, Suite 4600, New York, NY 10004. Phone 1-800-SPRINGER, fax (201) 348-4505, e-mail orders-ny@springer-sbm.com, or visit www.springeronline.com. Apress Media, LLC is a California LLC and the sole member (owner) is Springer Science + Business Media Finance Inc (SSBM Finance Inc). SSBM Finance Inc is a **Delaware** corporation.

For information on translations, please e-mail booktranslations@springernature.com; for reprint, paperback, or audio rights, please e-mail bookpermissions@springernature.com.

Apress titles may be purchased in bulk for academic, corporate, or promotional use. eBook versions and licenses are also available for most titles. For more information, reference our Print and eBook Bulk Sales web page at http://www.apress.com/bulk-sales.

Any source code or other supplementary material referenced by the author in this book is available to readers on GitHub at http://github.com/Apress/Generic-Data-Structures-and-Algorithms-in-Go.

Printed on acid-free paper

This book is dedicated to my wife Hanne.

Table of Contents

About the Author

Richard Wiener, PhD, authored or coauthored 22 professional, software development, and computer science textbooks published by Wiley, Addison-Wesley, Prentice Hall, Cambridge University Press, and Thompson. He served as founding Editor-in-Chief of the *Journal of Object-Oriented Programming* for 12 years and, later, founding Editor-in-Chief of the *Journal of Object Technology* for 9 years. He worked as Associate Professor of Computer Science at the University of Colorado Colorado Springs (UCCS) from 1977 to 2012. He served as Department Chair during the last four years at UCCS. He also served as consultant and software developer for IBM, HP, Boeing, Tektronix, DEC, and many other companies. He presented industry short courses all over the world from 1980 to 2006. He earned a BS and an MS in Electrical Engineering from the City University of New York and PhD from the Polytechnic Institute of New York.

About the Technical Reviewer

Fabio Claudio Ferracchiati is a senior consultant and a senior analyst/developer using Microsoft technologies. He works for Bluarancio (`www.bluarancio.com`). He is a Microsoft Certified Solution Developer for .NET, a Microsoft Certified Application Developer for .NET, a Microsoft Certified Professional, and a prolific author and technical reviewer. Over the past ten years, he's written articles for Italian and international magazines and coauthored more than ten books on a variety of computer topics.

Acknowledgments

The author thanks the reviewers at Apress including Gryffin Winkler and James Markham; the technical reviewer, Fabio Claudio Ferracchiati; and Steve Anglin for signing the book.

Introduction

This book is aimed at practicing Go software developers and students who wish to experience the excitement and see the benefits of data structures and algorithms in action. Because of its clean and readable syntax, outstanding support for concurrency and generics, and execution speed, Go was chosen to present the implementations of the data structures and algorithms along with many applications. It is assumed that the reader has basic familiarity with Go. The numerous code listings and their explanations will hopefully serve to improve your programming skills using Go.

The latest version of Go, Version 1.18, features genericity (generic and constrained generic parameters for data types and functions). This long-awaited addition to the Go language is ideally suited for use in building reusable data structure packages. Prior to Go Version 1.18, separate implementations of data structures and their associated algorithms were limited to a particular data type. So, for example, a list containing information of type **int** would have to be reimplemented if the underlying type was, for example, changed to **float64** or to a more complex user-defined custom type. With the new Version 1.18 of Go, generic and constrained generic data types remove this severe restriction. Generic and constrained generic data types will be featured throughout this book, and all source listings will use Version 1.18 of Go.

Computer science, like many sciences, has many areas of specialization – network security, e-commerce, general web application development, graphics, game design, database applications, encryption, natural language processing, text analysis, compiler design, operating systems, simulation, machine learning, and AI, just to name a few. Knowledge of the effective design and use of data structures and algorithms are useful in these areas of specialization and are therefore a fundamental part of computer science and software development methodology.

Over the years and because of application development in the areas mentioned and not mentioned previously, a consensus has emerged about which data structures and algorithms have the greatest utility. Nothing is static in this area, so new data structures and algorithms are being designed. The task of successfully advancing and moving ahead as a developer is greatly enhanced by studying the great works already established. The goal of education is learning how to learn. In the context of this book, the great works are the data structures and accompanying algorithms that have been shown to have utility in a large variety of computation problem domains.

This is not a theoretical book laden with formal proofs. There are many such books already available. For the data structures presented, not every use case is included. It is hoped and expected that after focusing on the major data structures and associated algorithms presented, the reader will be better prepared to extend their knowledge as they move forward in creating or discovering new data structures and algorithms. It is also hoped that by presenting a variety of problems that are solved by hitching a ride with one or more of the data structures and algorithms introduced, the reader will appreciate the power that mastery of this subject matter brings to the table of software development.

The use of concurrency in implementing data structures is a major feature of this book. Concurrent implementations are utilized whenever appropriate throughout the book.

Chapter 1 presents a tour of generics and concurrency in Go.

Chapters 2 through 16 present classic data structures and algorithms and show them in action. These include sorting and searching, stack, queue, lists, deque, hash table, binary search tree, AVL tree, red-black tree, heap, expression tree, and graph. Many examples and applications are presented. Dynamic programming and branch-and-bound algorithms are used to solve classic problems such as shortest path in a graph and minimum spanning tree.

Chapter 17 introduces combinatorial optimization problems and focuses on the famed Travelling Salesperson Problem (TSP). Exact solutions are intractable both in memory and execution time. A brute-force solution is presented in Chapter 17. A branch-and-bound solution is presented in Chapter 18. This sets the stage for Chapters 19 and 20, which present heuristic solutions to this problem.

Chapter 19 presents a simulated annealing heuristic solution, which is shown to be very effective in solving large TSP problems.

Chapter 20 presents a genetic algorithm, another effective heuristic solution to TSP.

Chapter 21 introduces machine learning and neural networks. A neural network is constructed from scratch and used to train a network to evaluate medical test results.

In summary, this book will

- Explore classical data structures and algorithms aimed at making your applications run faster or require less storage

- Use the new generic features of Go to build reusable data structures

- Utilize concurrency for maximizing application performance

- See the power of heuristic algorithms for computationally intractable problems

- Enhance and improve your Go programming skills

A Tour of Generics and Concurrency in Go

This chapter introduces the syntax and semantics of generics in Go. Many coding examples are presented that illustrate this new and powerful feature of Go. This sets the stage for the continued use of generics throughout the book.

Concurrency in Go is also reviewed in this chapter. Many coding examples are presented along with benchmarks that contrast the performance of algorithms with and without concurrency. This also sets the stage for the continued use of concurrency throughout the book.

1.1 Brief History and Description of Go

Go is a relatively new programming language released in late 2009 and developed at Google by Robert Griesemer (a Swiss computer scientist who helped create Google's V8 JavaScript engine), Rob Pike (a Canadian computer scientist and part of the Unix team at Bell Labs and creator of the Limbo programming language), and Ken Thompson (creator of the Unix operating system and the B programming language).

The Go programming language is sometimes called Golang. Why? The domain "go.org" wasn't available at the time the language was released, so golang.org (a mix of Go and language) was born. The official name of the language is Go, but the Twitter tag is #golang. Go figure!

One of the major goals in creating Go was to produce an easily readable, strong, and statically typed language with garbage collection and fast compilation and execution speed particularly suited for concurrent applications.

The **goroutine** is a lightweight process that requires less memory overhead than a normal thread seen in other languages such as Java and C#. A concurrent Go program may spawn thousands of goroutines running on a much smaller number of threads.

1

© Richard Wiener 2022
R. Wiener, *Generic Data Structures and Algorithms in Go*, https://doi.org/10.1007/978-1-4842-8191-8_1

The **channel** construct (to be explained later in this chapter) allows information to be passed into and taken out of goroutines and is used to synchronize these concurrent lightweight processes. Although parallel processing is not the primary objective of goroutines, they can be used to approximate this on a shared memory multiprocessor computer.

Go is a platform-independent language that runs on various Unix platforms including MacOS and also runs on MS Windows. Go applications compile to a binary executable so they can be distributed to a customer without having to package an interpreter and runtime libraries as is the case with Python and other interpreted languages.

Go, like many recent languages, is a public open source project. There are a bevy of free tools that are downloadable. New packages are constantly being released, so much of the power of the language resides outside the language in the plethora of high-quality packages available to the programmer. In this sense, Go is like Python.

Among the tools that are available are high-quality editors, debuggers, and IDEs. Go requires a prescribed format, so the **gofmt** tool is often integrated into various code editors. Having a standard code format provides a huge advantage to Go programming teams as well as solo programmers inspecting the code written by others.

So what is missing in Go? What is its downside? Up until the most recent and perhaps most important new release, Version 1.18, Go lacked genericity. With this new release of Go, this major shortcoming is gone.

Now one can build an algorithm or data structure that does not have to be modified every time the underlying information to be stored changes. Data structure and algorithm implementations can focus on the core logic needed to manipulate the information. A new syntax associated with generics allows a programmer to precisely describe the requirements that data must satisfy to be stored in a particular data structure. This furthers a programmer's ability to have a program specify its intent in the code itself. The use of constrained and unconstrained generic parameters is introduced and illustrated in the next section.

1.2 Introducing Generic Parameters

In this section, we present a series of examples that introduce and illustrate the use of generic-type parameters, both unconstrained and constrained.

In the first several code listings, we present a set of related problems of adding a new student to an existing slice of students. First, we add just the name of the student to our existing slice. Next, we add the student's ID number to a slice containing ID numbers. Next, we add a struct containing name, ID, and age to an existing slice of structs. Then finally, we bring generics on stage and show the simplification that is achievable using a generic-type parameter.

Adding a New Student by Name

Consider the simple Go application given in Listing 1-1.

Listing 1-1. A slice of students

```go
package main

import(
    "fmt"
)

func addStudent(students []string, student string) []string {
    return append(students, student)
}

func main() {
    students := []string{} // Empty slice
    result := addStudent(students, "Michael")
    result = addStudent(result, "Jennifer")
    result = addStudent(result, "Elaine")
    fmt.Println(result)
}
/*
Output:
[Michael Jennifer Elaine]
*/
```

The function *addStudent* takes a slice of string representing the current collection of students as the first parameter and a string representing a new student to be added to the collection as the second parameter. The *append* function is used to add the new student to the existing slice, and that result is returned.

Adding a New Student by ID Number

Suppose we wish to specify the slice of students using their ID number, an int, instead of their name, a string.

We would need to modify Listing 1-1 as shown in Listing 1-2.

Listing 1-2. Adding student IDs

```go
package main

import(
    "fmt"
)

func addStudent(students []string, student string) []string {
    return append(students, student)
}

func addStudentID(students []int, student int) []int {
    return append(students, student)
}

func main() {
    students := []string{} // Empty slice
    result := addStudent(students, "Michael")
    result = addStudent(result, "Jennifer")
    result = addStudent(result, "Elaine")
    fmt.Println(result)

    students1 := []int{} // Empty slice
    result1 := addStudentID(students1, 155)
    result1 = addStudentID(result1, 112)
    result1 = addStudentID(result1, 120)
    fmt.Println(result1)
}
/* Output
[Michael Jennifer Elaine]
[155 112 120]
*/
```

The logic in function ***addStudentID*** is essentially the same as in function ***addStudent***. Only the base type of the slice is changed from **string** to **int**.

Adding a New Student by Student Struct

And to take this one step further, suppose we define a ***Student*** type as

```
type Student struct {
    Name string
    ID int
    age float64
}
```

and we modify Listing 1-2 as shown in Listing 1-3.

Listing 1-3. Adding Student type to the mix

```
package main

import(
    "fmt"
)

type Student struct {
    Name string
    ID int
    age float64
}

func addStudent(students []string, student string) []string {
    return append(students, student)
}

func addStudentID(students []int, student int)  []int {
    return append(students, student)
}
```

5

```go
func addStudentStruct(students []Student, student Student) []Student {
    return append(students, student)
}
func main() {
    students := []string{} // Empty slice
    result := addStudent(students, "Michael")
    result = addStudent(result, "Jennifer")
    result = addStudent(result, "Elaine")
    fmt.Println(result)

    students1 := []int{} // Empty slice
    result1 := addStudentID(students1, 155)
    result1 = addStudentID(result1, 112)
    result1 = addStudentID(result1, 120)
    fmt.Println(result1)

    students2 := []Student{} // Empty slice
    result2 := addStudentStruct(students2, Student{"John", 213, 17.5} )
    result2 = addStudentStruct(result2,  Student{"James", 111, 18.75} )
    result2 = addStudentStruct(result2,  Student{"Marsha", 110, 16.25} )
    fmt.Println(result2)
}
/* Output
[Michael Jennifer Elaine]
[155 112 120]
[{John 213 17.5} {James 111 18.75} {Marsha 110 16.25}]
*/
```

Having to add a new function each time we wish to add a new underlying data type to our various student collections is tedious and a major downside to earlier versions of Go.

Introducing Generics

Enter Go, Version 1.18, that introduces support for generics.

A generic solution to this problem is presented in Listing 1-4.

Listing 1-4. Generic solution to problem

```go
package main

import (
    "fmt"
)

type Stringer = interface {
    String() string
}

type Integer int

func (i Integer) String() string {
    return fmt.Sprintf("%d", i)
}

type String string

func (s String) String() string {
    return string(s)
}

type Student struct {
    Name string
    ID int
    Age float64
}

func (s Student) String() string {
    return fmt.Sprintf("%s %d %0.2f", s.Name, s.ID, s.Age)
}

func addStudent[T Stringer](students []T, student T) []T {
    return append(students, student)
}

func main() {
    students := []String{}
    result := addStudent[String](students, "Michael")
```

```
    result = addStudent[String](result, "Jennifer")
    result = addStudent[String](result, "Elaine")
    fmt.Println(result)

    students1 := []Integer{}
    result1 := addStudent[Integer](students1, 45)
    result1 = addStudent[Integer](result1, 64)
    result1 = addStudent[Integer](result1, 78)
    fmt.Println(result1)

    students2 := []Student{}
    result2 := addStudent[Student](students2, Student{"John", 213, 17.5} )
    result2 = addStudent[Student](result2, Student{"James", 111, 18.75} )
    result2 = addStudent(result2,  Student{"Marsha", 110, 16.25} )
    fmt.Println(result2)
}
/* Output
[Michael Jennifer Elaine]
[45 64 78]
[John 213 17.50 James 111 18.75 Marsha 110 16.25]
*/
```

Stringer Type

A type **Stringer** is defined as an interface containing a single method signature:

String() string.

Any entity that implements this type by having a well-defined **String()** definition can be converted to a string for output purposes. Since we wish to be able to output our various student collections (slices), we constrain the data type, *T*, in the signature of our generic **addStudent** function to be of type **Stringer**.

Constrained Generic Type

The generic signature of our single **addStudent** function becomes

func addStudent[T Stringer](students []T, student T) [ ]T

Types **Integer**, **String**, and **Student** are defined along with their definitions of **String()** so that we can use generic function **addStudent** using each of these **Stringer** types.

8

Implementing an Interface

In Go, one implements an interface implicitly by implementing the function(s) specified in the interface definition. In this case, any type that implements a **String()** function can be considered to be of type *Stringer*.

Instantiating a Generic Type

In main, after declaring students to be an empty slice of *String* (not a slice of string), we invoke **addStudent[String](students, "Michael").**

The generic parameter *T* constrained to be of type *Stringer* is replaced by the actual instantiated type *String* which we know is of type **Stringer** because it implements the *Stringer* interface (which has a concrete definition of *String()*).

We next use *addStudent* with *Integer* used as the *Stringer* type. And finally, we use *addStudent* with *Student* as the *Stringer* type.

Unconstrained Generic Type any

The Go compiler can do type inferencing if we replace the constrained generic type **[T Stringer]** with the unconstrained type **any**.

Listing 1-5 presents a simpler, less verbose, generic implementation of *addStudent* along with a driver function *main* that exercises this generic function.

Listing 1-5. Simpler generic function addStudent

```
package main

import (
    "fmt"
)

type Student struct {
    Name string
    ID int
    Age float64
}
```

```go
func addStudent[T any](students []T, student T) []T {
    return append(students, student)
}

func main() {
    students := []string{}
    result := addStudent[string](students, "Michael")
    result = addStudent[string](result, "Jennifer")
    result = addStudent[string](result, "Elaine")
    fmt.Println(result)

    students1 := []int{}
    result1 := addStudent[int](students1, 45)
    result1 = addStudent[int](result1, 64)
    result1 = addStudent[int](result1, 78)
    fmt.Println(result1)

    students2 := []Student{}
    result2 := addStudent[Student](students2, Student{"John", 213, 17.5} )
    result2 = addStudent[Student](result2, Student{"James", 111, 18.75} )
    result2 = addStudent(result2,  Student{"Marsha", 110, 16.25} )
    fmt.Println(result2)
}
/* Output
[Michael Jennifer Elaine]
[45 64 78]
[John 213 17.50 James 111 18.75 Marsha 110 16.25]
*/
```

Using type inferences, the compiler uses the default conversions of **string**, **int**, and **Student** to allow program output by converting each of these types to string.

Benefits of Generics

In Listing 1-5, the benefits of generics are evident. The simple algorithm for appending a new student (second parameter) to the existing collection of students is expressed independently of the type being used to represent a student.

Suppose we wish to sort each collection of students prior to outputting the collection. We can do so with the sort package discussed in the next section.

Using Go's Sort Package

The **Sort** function in Go's **sort** package, **sort.Sort**, requires that the type in the slice being sorted must implement three methods:

1. **Len**

2. **Less**

3. **Swap**

We show how a generic collection implemented as a slice can be sorted.

We define *OrderedSlice* as follows and provide the required group of *Len*, *Less*, and *Swap*.

```go
// Group of functions that ensure that an OrderedSlice can be sorted
type OrderedSlice[T Ordered] []T // T must implement < and >

func (s OrderedSlice[T]) Len() int {
    return len(s)
}

func (s OrderedSlice[T]) Less(i, j int) bool {
    return s[i] < s[j]
}

func (s OrderedSlice[T]) Swap(i, j int) {
    s[i], s[j] = s[j], s[i]
}
// end group for OrderedSice
```

Sort Type

We introduce another type, *SortType*, along with the required group of *Len*, *Less*, and *Swap*.

```go
// Group of functions that ensure that SortType can be sorted
type SortType[T any] struct {
    slice []T
    compare func(T, T) bool
}

func (s SortType[T]) Len() int {
    return len(s.slice)
}

func (s SortType[T]) Less(i, j int) bool {
    return s.compare(s.slice[i], s.slice[j])
}

func (s SortType[T]) Swap(i, j int) {
    s.slice[i], s.slice[j] = s.slice[j], s.slice[i]
}
// end of group for SortType
```

Finally, we define a function, *PerformSort*, that uses *SortType* as follows:

```go
func PerformSort[T any](slice []T, compare func(T, T) bool) {
    sort.Sort(SortType[T]{slice, compare})
}
```

The user of *PerformSort* must supply a function for comparing two elements of type T.

Listing 1-6 integrates this functionality into the code that implements the generic *addStudent* function to allow us to use the imported **sort** package and its function **Sort**.

Listing 1-6. Building and sorting slices of students

```go
package main

import (
    "fmt"
    "sort"
)
```

```go
type Ordered interface {
        ~int | ~float64 | ~string
}

type Student struct {
    Name string
    ID int
    Age float64
}

func addStudent[T any](students []T, student T) []T {
    return append(students, student)
}

// Group of functions that ensure that an OrderedSlice can be sorted
type OrderedSlice[T Ordered] []T // T must implement < and >

func (s OrderedSlice[T]) Len() int {
    return len(s)
}

func (s OrderedSlice[T]) Less(i, j int) bool {
    return s[i] < s[j]
}

func (s OrderedSlice[T]) Swap(i, j int) {
    s[i], s[j] = s[j], s[i]
}
// end group for OrderedSice

// Group of functions that ensure that SortType can be sorted
type SortType[T any] struct {
    slice []T
    compare func(T, T) bool
}
func (s SortType[T]) Len() int {
    return len(s.slice)
}
```

```go
func (s SortType[T]) Less(i, j int) bool {
    return s.compare(s.slice[i], s.slice[j])
}

func (s SortType[T]) Swap(i, j int) {
    s.slice[i], s.slice[j] = s.slice[j], s.slice[i]
}
// end of group for SortType
func PerformSort[T any](slice []T, compare func(T, T) bool) {
    sort.Sort(SortType[T]{slice, compare})
}

func main() {
    students := []string{}
    result := addStudent[string](students, "Michael")
    result = addStudent[string](result, "Jennifer")
    result = addStudent[string](result, "Elaine")
    sort.Sort(OrderedSlice[string](result))
    fmt.Println(result)

    students1 := []int{}
    result1 := addStudent[int](students1, 78)
    result1 = addStudent[int](result1, 64)
    result1 = addStudent[int](result1, 45)
    sort.Sort(OrderedSlice[int](result1))
    fmt.Println(result1)

    students2 := []Student{}
    result2 := addStudent[Student](students2, Student{"John", 213, 17.5} )
    result2 = addStudent[Student](result2, Student{"James", 111, 18.75} )
    result2 = addStudent(result2,   Student{"Marsha", 110, 16.25} )
    PerformSort[Student](result2, func(s1, s2 Student) bool {
        return s1.Age < s2.Age // comparing two Student values
    })
    fmt.Println(result2)
}
```

```
/* Output
[Elaine Jennifer Michael]
[45 64 78]
[{Marsha 110 16.25} {John 213 17.5} {James 111 18.75}]
*/
```

Map Functions

Map functions in Go are commonplace and perform a transformation in a slice to produce a new slice with the transformed results. Consider this example:

```go
func MyMap(input []int, f func(int) int) []int {
    result := make([]int, len(input))
    for index, value := range input {
        result[index] = f(value)
    }
    return result
}

func main() {
    slice := []int{1, 5, 2, 7, 4}
    result := MyMap(slice, func(i int) int {
        return i * i
    })
    fmt.Println(result)
}
/* Output
[1 25 4 49 16]
*/
```

The *MyMap* function produces an output slice containing the square of the integers contained in the input slice. After declaring *result* to be a slice of **len(slice)** integers, it iterates over the range of values in *input*, transforming each value based on the function *f* passed in to *MyMap*.

Making MyMap Generic

MyMap can be made generic as follows:

```go
func GenericMap[T1, T2 any](input []T1, f func(T1) T2) []T2 {
  result := make([]T2, len(input))
  for index, value := range input {
    result[index] = f(value)
  }
  return result
}
```

Function ***GenericMap*** takes two generic parameters, ***T1*** and ***T2***. Using the function *f* that is passed in, it transforms the data in the ***input*** slice to type ***T2***. Here, **T1** and **T2** are not constrained. They are of type **any**.

Filter Functions

Filter functions in Go are also commonplace and perform a filtering operation on an input slice based on a function passed in. Consider this example:

```go
func MyFilter(input []float64, f func(float64) bool) []float64 {
    var result []float64
    for _, value := range input {
        if f(value) == true {
            result = append(result, value)
        }
    }
    return result
}

func main() {
    input := []float64{17.3, 11.1, 9.9, 4.3, 12.6}
    res := MyFilter(input, func(num float64) bool {
        if num <= 10.0 {
            return true
        }
```

CHAPTER 1   A TOUR OF GENERICS AND CONCURRENCY IN GO

```
        return false
    })
    fmt.Println(res)
}
/* Output
[9.9 4.3]
*/
```

Here, any value in the input slice that is less than or equal to 10.0 is retained, and all other values are filtered out.

Making MyFilter Generic

MyFilter can be made generic as follows:

```
func GenericFilter[T any](input []T, f func(T) bool) []T {
  var result []T
  for _, val := range input {
    if f(val) {
     result = append(result, val)
    }
  }
  return result
}
```

In Listing 1-7, we exercise the generic map and filter functions.

Listing 1-7. Using generic map and filter functions

```
package main

import (
    "fmt"
)
```

```go
func GenericMap[T1, T2 any](input []T1, f func(T1) T2) []T2 {
  result := make([]T2, len(input))
  for index, value := range input {
    result[index] = f(value)
  }
  return result
}

func GenericFilter[T any](input []T, f func(T) bool) []T {
  var result []T
  for _, val := range input {
    if f(val) {
     result = append(result, val)
    }
  }
  return result
}

func main() {
    input := []float64{-5.0, -2.0, 4.0, 8.0}
    result1 := GenericMap[float64, float64](input, func(n float64)
    float64 {
        return n * n
    })
    fmt.Println(result1)

    greaterThanFive := GenericFilter[int]([]int{4, 6, 5, 2, 20, 1, 7},
        func(i int) bool {
            return i > 5
        })
    fmt.Println(greaterThanFive)

    oddNumbers := GenericFilter[int]([]int{4, 6, 5, 2, 20, 1, 7},
        func(i int) bool {
            return i % 2 == 1
        })
    fmt.Println(oddNumbers)
```

```
lengthGreaterThan3 := GenericFilter[string]([]string{"hello", "or",
"the", "maybe"}, func(s string) bool {
                            return len(s) > 3
                    })
    fmt.Println(lengthGreaterThan3)
}
/* Output
[25 4 16 64]
[6 20 7]
[5 1 7]
[hello maybe]
*/
```

We now turn our attention to the use of concurrency.

1.3 Concurrency

Concurrency allows a program to process multiple tasks at the same time (parallel processing where each task is assigned to a separate processor) or what appears to be at the same time where tasks are multiplexed so progress is made on all tasks over time. If the multiplexing is very fast, it appears that the concurrent processes are running at the same time but are run in overlapping periods of time.

In most languages that support concurrent processing, the **thread** construct is used to support concurrency. There is memory overhead associated with a thread, so the number of threads that can be spawned at the same time is limited.

Goroutine

In developing the Go language, Google introduced a lightweight process called a **goroutine** that requires less memory overhead than a thread. Their motivation was to be able to serve multiple HTML web pages made from many web browsers at the same time.

Goroutines are functions that run concurrent with other functions. When a regular function is invoked, the code below the function gets executed after the function completes its work. When a goroutine function is invoked, the code directly below it gets executed immediately since the goroutine runs concurrently with code beneath it.

19

We illustrate this in Listing 1-8.

Listing 1-8. Simple goroutine running concurrent with main

```go
package main

import (
    "fmt"
    "time"
)
func regularFunction() {
    fmt.Println("Just entered regularFunction()")
    time.Sleep(5 * time.Second)

}

func goroutineFunction() {
    fmt.Println("Just entered goroutineFunction()")
    time.Sleep(10 * time.Second)
    fmt.Println("goroutineFunction finished its work")
}

func main() {
    go goroutineFunction()
    fmt.Println("In main one line below goroutineFunction()")
    regularFunction()
    fmt.Println("In main one line below regularFunction()")
}
/* Output
In main, one line below goroutineFunction()
Just entered regularFunction()
Just entered goroutineFunction()
In main one line below regularFunction()
*/
```

When the ***goroutineFunction*** is launched as a goroutine using ***go goroutineFunction()***, it runs concurrently with the ***main*** function, which is a goroutine. The first line of output occurs immediately even though the ***goroutineFunction*** requires ten seconds to complete its work. When the ***regularFunction()*** is invoked next, five seconds elapses before the line of output. "In main, one line below regularFunction()" is emitted. Function main terminates immediately after this output is emitted, which ends the program before the goroutineFunction can complete its work. It gets interrupted and terminates when the program ends.

If we swap the time delays so that the ***goroutineFunction*** has a time delay of five seconds and the ***regularFunction*** has a time delay of ten seconds, the output becomes

```
In main one line below goroutineFunction()
Just entered regularFunction()
Just entered goroutineFunction()
goroutineFunction finished its work
In main, one line below regularFunction()
```

Now the goroutine running concurrently with main completes it work before the ***regularFunction*** and before the ***main*** goroutine exits.

WaitGroup

Go provides a mechanism for allowing multiple goroutines to all complete their work before main exits while killing off unfinished goroutines.

We introduce the sync package and the **WaitGroup** construct and illustrate its use in Listing 1-9.

Listing 1-9. The sync package and WaitGroup

```
package main

import (
    "fmt"
    "time"
    "math/rand"
    "sync"
)

var wg sync.WaitGroup

func outputStrings() {
    defer wg.Done()
    strings := [5]string{"One", "Two", "Three", "Four", "Five"}
```

```go
    for i := 0; i < 5; i++ {
        delay := 1 + rand.Intn(3)
        time.Sleep(time.Duration(delay) * time.Second)
        fmt.Println(strings[i])
    }
}

func outputInts() {
    defer wg.Done()
    for i := 0; i < 5; i++ {
        delay := 1 + rand.Intn(3)
        time.Sleep(time.Duration(delay) * time.Second)
        fmt.Println(i)
    }
}

func outputFloats() {
    defer wg.Done()
    for i := 0; i < 5; i++ {
        delay := 1 + rand.Intn(3)
        time.Sleep(time.Duration(delay) * time.Second)
        fmt.Println(float64(i * i) + 0.5)
    }
}

func main() {
    wg.Add(3)
    go outputStrings()
    go outputInts()
    go outputFloats()
    wg.Wait()
}
/* Output
One
0.5
0
1
```

```
Two
1.5
2
4.5
3
Three
Four
4
9.5
Five
16.5
*/
```

This program does nothing useful except illustrating the **WaitGroup** construct and shows three goroutines running concurrently.

A global variable **wg** of type **sync.WaitGroup** is declared.

In **main**, we invoke **wg.Add(3)**. The last line of code in main is **wg.Wait()**. This causes main to pause until the value in **wg** is zero. This assures us that all three goroutines complete their work before the program terminates.

In each of the goroutines, the first line of code, **defer wg.Done()**, causes the value of the global variable **wg** to be decremented when the goroutine completes its work. When **wg** reaches a value of zero, the function **main** is allowed to exit.

The sequence of random numbers generated is the same each time the program is run, but the output sequence varies from run to run. This is because the time multiplexer allocates different chunks of execution time to each concurrent goroutine differently each time the program runs. After the second run of the program, the output is

```
One
0
0.5
1.5
Two
1
4.5
2
Three
9.5
```

```
16.5
Four
3
Five
4
```

The Channel

We often want to be able to synchronize the sequence of goroutines and have them communicate with each other. We introduce the powerful construct of the **channel** to accomplish this.

Consider the goroutines in Listing 1-10.

Listing 1-10. Deadlock

```go
package main

import (
    "fmt"
    "time"
    "sync"
)

var wg sync.WaitGroup

func pingGenerator(c chan string) {
    defer wg.Done()
    for i := 0; i < 5; i++ {
        c <- "ping"
        time.Sleep(time.Second * 1)
    }
}

func output(c chan string) {
    defer wg.Done()
    for {
```

```
        value := <- c
        fmt.Println(value)
    }
}
func main() {
    c := make(chan string)

    wg.Add(2)
    go pingGenerator(c)
    go output(c)

    wg.Wait()
}
/* Output
ping
ping
ping
ping
ping
fatal error: all goroutines are asleep - deadlock!
*/
```

The first line of code in main initializes a channel, *c*. Channels must be initialized with a ***make*** statement before they can be used.

As in the previous listing, we create a ***WaitGroup*** variable, ***wg***, with the initial value of 2.

We next launch the two goroutines, ***pingGenerator*** and ***output***, passing the channel variable *c* to each.

The ***pingGenerator*** goroutine assigns the string "ping" to the channel variable *c* every second and does this five times. The left arrow from the value "ping" to the variable *c* represents the assignment of the "ping" value to *c*.

The channel must be empty for this assignment to be made. In the ***output*** goroutine, the assignment to ***value***, using ***value :=*** <- *c*, gobbles up the channel variable *c* as soon as it is assigned in the ***pingGenerator***. This occurs every second. During the time between "ping" assignments from the ***pingGenerator***, the ***value*** assignment is blocked. That is execution is halted within the output goroutine until there is information in the channel assigning to value. So the two goroutines are being affected by the channel variable *c*, common to both.

There is a problem. When the ***pingGenerator*** has emitted its five "ping" assignments, each displayed on the console through the ***output*** goroutine, it blocks while waiting for a sixth channel assignment. This never occurs. The program crashes with the error message shown earlier. A deadlock has occurred. The program cannot terminate.

Select Statement

We can resolve this problem by modifying the ***output*** goroutine and using a **select** statement.

```
func output(c chan string) {
      for {
      select {
         case value := <- c:
             fmt.Println(value)
         case <-time.After(3 * time.Second):
             fmt.Println("Program timed out.")
             wg.Done()
         }
    }
}
```

In a **select** statement, the case that occurs first gets executed. If two or more cases are ready to execute, the system chooses one at random. Since the channel ***c*** gets assigned to ***value*** every second (blocks between assignments), the program outputs the five "ping" assignments. Instead of deadlocking as before, the second case gets executed after three seconds from the time the fifth and final "ping" is assigned to ***value***.

Use a quit Channel to Avoid Using WaitGroup

We can use a ***quit*** channel to block ***main*** from exiting and avoid the use of **WaitGroup** as shown in Listing 1-11.

Listing 1-11. Using a quit channel instead of WaitGroup

```go
package main

import (
    "fmt"
    "time"
)

var quit chan bool

func pingGenerator(c chan string) {
    for i := 0; i < 5; i++ {
        c <- "ping"
        time.Sleep(time.Second * 1)
    }
}

func output(c chan string) {
    for {
        select {
            case value := <- c:
                fmt.Println(value)
            case <-time.After(3 * time.Second):
                fmt.Println("Program timed out.")
                quit <- true
        }
    }
}

func main() {
    quit = make(chan bool)
    c := make(chan string)
    go pingGenerator(c)
    go output(c)
    <- quit
}
```

```
/* Output
ping
ping
ping
ping
ping
Program timed out.
*/
```

The ***quit*** channel is initialized as the first line of code in ***main***. The last line of code in ***main***, **<- quit**, blocks ***main*** from ending until a Boolean value is assigned to ***quit***. This occurs in the second case statement in goroutine ***output***.

This mechanism for controlling the end of the program is simpler and less encumbered than using **WaitGroup**.

We add the inevitable ***pongGenerator*** to this program.

Channel Direction

Channel direction can be added to a goroutine signature as shown in Listing 1-12. An arrow pointing to the ***chan*** from the right, as shown in the signatures to ***pingGenerator*** and ***pongGenerator***, requires the goroutine to assign to the channel (a generator). An arrow to the left of ***chan*** and pointing to the channel variable requires the goroutine to only consume values in the channel.

If an attempt is made to send information to the channel when it is specified as a consumer, or if an attempt is made to access information from the channel in the case that it is specified as a generator, a compiler error will occur.

Listing 1-12. Ping pong using direction channels in goroutine signatures

```
package main

import (
    "fmt"
    "time"
)
var quit chan bool
```

```go
func pingGenerator(c chan<- string) {
    // The channel can only be sent to - a generator
    for i := 0; i < 5; i++ {
    c <- "ping"
    }
}

func pongGenerator(c chan<- string) {
    // Information can only be sent to the channel - a generator
    for i := 0; i < 5; i++ {
        c <- "pong"
    }
}

func output(c <- chan string) {
    // Information can only be received from the channel - a consumer
        for {
        time.Sleep(time.Second * 1)
        select {
           case value := <- c:
               fmt.Println(value)
           case <-time.After(3 * time.Second):
               fmt.Println("Program timed out.")
               quit <- true
        }
    }
}

func main() {
    quit = make(chan bool)
    c := make(chan string)
    go pingGenerator(c)
    go pongGenerator(c)
    go output(c)
    <- quit
}
```

```
/* Output
ping
pong
ping
pong
ping
pong
ping
pong
ping
pong
Program timed out.
*/
```

We have moved the one-second time delay into the ***output*** goroutine. This allows the ping and pong generators to alternate since each assignment to the channel blocks alternately until the channel is read by the consuming ***output*** goroutine.

Race Condition

A pervasive problem using concurrency is race condition. This problem occurs when two or more goroutines modify the same shared data.

A simple example is presented in Listing 1-13.

Listing 1-13. Example of race condition

```
package main

import (
    "fmt"
    "sync"
)

const (
    number = 1000
)
```

```go
var countValue int

func main() {
    var wg sync.WaitGroup
    wg.Add(number)

    for i := 0; i < number; i++ {
        go func() {
            countValue++
            wg.Done()
        }()
    }
    wg.Wait()
    fmt.Printf("\ncountValue = %d", countValue)
}
```

One thousand goroutines are spawned in a for-loop within **main**. Each goroutine increments **countValue** exactly once. Therefore, one would expect the output of the program to be 1000.

Each time the program is run, the output is different. This is because of the conflict of multiple goroutines attempting to modify the global **changeValue** at nearly the same time. There is no error message generated by the system. But the output is incorrect.

Mutex

We can correct the race-condition problem by using a mutex. This locks the global **countValue** while each goroutine modifies its value and protects this shared data from being corrupted.

Listing 1-14 shows the use of a **mutex** to remove the race condition.

Listing 1-14. Using mutex to avoid race condition

```go
package main

import (
    "fmt"
    "sync"
)
```

```go
const number = 1000

var countValue int
var m sync.Mutex

func main() {
    var wg sync.WaitGroup
    wg.Add(number)
    for i := 0; i < number; i++ {
        go func() {
            m.Lock()
            countValue++
            m.Unlock()
            wg.Done()
        }()
    }
    wg.Wait()
    fmt.Printf("\ncountValue = %d", countValue)
}
```

The code *m.Lock()* within each goroutine protects the global *countValue* from modification outside of the goroutine in which it is invoked. No other goroutine can change *countValue* until the *m.Unlock()* is invoked.

Program execution is slowed down using the **mutex**, but the program is protected from the race condition shown in Listing 1-13.

Playing Chess Using Goroutines

Listing 1-15 simulates the sequence of two chess players making moves using goroutines.

Listing 1-15. Concurrent moves in chess

```go
// This sample program demonstrates how to use an unbuffered
// channel to simulate a move of chess between two goroutines.
package main
```

```go
import (
    "fmt"
    "math/rand"
    "time"
)

var quit chan bool

func main() {
    rand.Seed(time.Now().UnixNano())
    move := make(chan int)
    quit = make(chan bool)

    // Launch two players.
    go player("Bobby Fischer", move)
    go player("Boris Spassky", move)

    // Start the move
    move <- 1
    <-quit // Blocks until quit assigned a value
}

// player simulates a person moving in chess.
func player(name string, move chan int) {
    // This function takes data out of the move channel
    // and puts data back into the move channel
    for {
        // Wait for turn to play
        turn := <-move // blocks until move assigned a value (every second)

        // Pick a random number and see if we lose the move
        n := rand.Intn(100)
        if n <= 5 && turn >= 5 {
            fmt.Printf("Player %s was check mated and loses!", name)
            quit <- true
            return
        }
```

```
        // Display and then increment the total move count by one.
        fmt.Printf("Player %s has moved. Turn %d.\n", name, turn)
        turn++

        // Yield the turn back to the opposing player
        time.Sleep(1 * time.Second)
        move <- turn
    }
}
/*
Player Boris Spassky has moved. Turn 1.
Player Bobby Fischer has moved. Turn 2.
Player Boris Spassky has moved. Turn 3.
Player Bobby Fischer has moved. Turn 4.
Player Boris Spassky has moved. Turn 5.
Player Bobby Fischer has moved. Turn 6.
Player Boris Spassky has moved. Turn 7.
Player Bobby Fischer has moved. Turn 8.
Player Boris Spassky has moved. Turn 9.
Player Bobby Fischer has moved. Turn 10.
Player Boris Spassky has moved. Turn 11.
Player Bobby Fischer has moved. Turn 12.
Player Boris Spassky has moved. Turn 13.
Player Bobby Fischer has moved. Turn 14.
Player Boris Spassky has moved. Turn 15.
Player Bobby Fischer has moved. Turn 16.
Player Boris Spassky has moved. Turn 17.
Player Bobby Fischer has moved. Turn 18.
Player Boris Spassky has moved. Turn 19.
Player Bobby Fischer has moved. Turn 20.
Player Boris Spassky has moved. Turn 21.
Player Bobby Fischer has moved. Turn 22.
Player Boris Spassky has moved. Turn 23.
Player Bobby Fischer has moved. Turn 24.
Player Boris Spassky has moved. Turn 25.
Player Bobby Fischer was check mated and loses!
*/
```

The first line within the for-loop of the goroutine *player* blocks until an int is taken out of the *move* channel. With a 5 percent probability, the player loses the game and sets *quit* to true.

After outputting that the player has moved and posting the player's turn, it puts an int back into the *move* channel, freeing the other player to move.

Fibonacci Numbers Using Goroutines

The next example, in Listing 1-16, shows how we can output a sequence of Fibonacci numbers using a goroutine.

Listing 1-16. Fibonacci numbers using a goroutine

```go
package main

import "fmt"

func fibonacci(c chan<- int, quit <-chan bool) {
    x, y := 0, 1
    for {
        select {
        case c <- x:
            x, y = y, x + y // Generates the sequence
        case <- quit:
            fmt.Println("quit")
            return
        }
    }
}

func main() {
    c := make(chan int)
    quit := make(chan bool)
    go func() {
        for i := 0; i < 20; i++ {
            fmt.Println(<-c)
        }
        quit <- true
```

```
    }()
    fibonacci(c, quit)
}
/* Output
0
1
1
2
3
5
8
13
21
34
55
89
144
233
377
610
987
1597
2584
4181
quit
*/
```

The first parameter, *c*, in function *fibonacci* puts information into the channel, and the second parameter, *quit*, takes information out of the channel.

The goroutine is launched within *main* as an internal function. The *Println(<-c)* statement blocks until the *fibonacci* function puts the value **x** into the integer channel **c**.

The **select** statement in function **fibonacci** either takes the next value of **x** into channel **c** or ends the program as soon as *quit* becomes true. The actual *fibonacci* sequence is computed as the second line within the *case c <-x* statement.

In the next section, we examine the possible performance improvements that are attainable by using concurrency.

1.4 Benchmarking Concurrent Applications

The goal in using concurrency is to speed up program execution. There is overhead in deploying goroutines, so sometimes, using concurrency is counterproductive. Because debugging concurrent code is challenging and dealing with deadlocks and race conditions is sometimes tricky, one needs to be careful when crafting concurrent solutions to a problem. Testing a concurrent application and comparing its performance with a nonconcurrent solution is useful.

In this section, we present several applications in which we compare the performance of a concurrent solution with a nonconcurrent solution. Since the result of a benchmark test is dependent on the processor and memory used, the ambient workload of the machine (how many processes are running in the background), and the machine architecture, one must be careful in generalizing and possibly drawing incorrect conclusions from a benchmark result.

Consider the program in Listing 1-17. Here, we compare the time required to construct and sum 100 million floating-point numbers into a slice with and without concurrency.

Listing 1-17. Computing Sum With and Without Concurrency

```go
package main

import "fmt"
import "time"
import "sync"

var output1 float64
var output2 float64
var output3 float64
var output4 float64

var wg sync.WaitGroup

func worker1() {
    defer wg.Done()
    var output []float64
    sum := 0.0
    for index := 0; index < 100_000_000; index++ {
```

```go
        output = append(output, 89.6)
        sum += 89.6
    }
    output1 = sum
}
func worker2() {
    defer wg.Done()
    var output []float64
    sum := 0.0
    for index := 0; index < 100_000_000; index++ {
        output = append(output, 64.8)
        sum += 64.8
    }
    output2 = sum
}

func worker3() {
    defer wg.Done()
    var output []float64
    sum := 0.0
    for index := 0; index < 100_000_000; index++ {
        output = append(output, 956.8)
        sum += 956.8
    }
    output3 = sum
}

func worker4() {
    defer wg.Done()
    var output []float64
    sum := 0.0
    for index := 0; index < 100_000_000; index++ {
        output = append(output, 1235.8)
        sum += 1235.8
    }
    output4 = sum
}
```

```go
func main() {
    wg.Add(8)
    // Compute time with no concurrent processing
    start := time.Now()
    worker1()
    worker2()
    worker3()
    worker4()
    elapsed := time.Since(start)
    fmt.Println("\nTime for 4 workers in series: ", elapsed)
    fmt.Printf("Output1: %f \nOutput2: %f  \nOutput3: %f  \nOutput4: %f\n",
                output1, output2, output3, output4)

    // Compute time with concurrent processing
    start = time.Now()
    go worker1()
    go worker2()
    go worker3()
    go worker4()
    wg.Wait()
    elapsed = time.Since(start)
    fmt.Println("\nTime for 4 workers in parallel: ", elapsed)
    fmt.Printf("Output1: %f \nOutput2: %f  \nOutput3: %f  \nOutput4: %f",
                output1, output2, output3, output4)
}

/* Output on a Macbook Pro with M1 Max chip with 10-core CPU, 32-core GPU,
and 16-core Neural Engine
Time for 4 workers in series:   1.133640541s
Output1: 8960000016.634367
Output2: 6480000011.637030
Output3: 95680000176.244049
Output4: 123580000205.352280

Time for 4 workers in parallel:   756.305958ms
Output1: 8960000016.634367
Output2: 6480000011.637030
```

```
Output3: 95680000176.244049
Output4: 123580000205.352280
*/
```

Each worker function appends a **float64** value to construct an output slice of 100 million values while computing the sum in the slice.

In main, we compute and output the computation time if the worker functions are executed sequentially. Then we execute the four worker functions concurrently using goroutines. We compare the computation time and verify the correctness of the results by outputting the sums with and without concurrency.

The computation time running the four worker functions concurrently is 67 percent the time running the four functions sequentially. As expected, the sums computed are the same.

Suppose we wish to speed up the computation of summing a sequence of numbers by using concurrency. Listing 1-18 demonstrates this.

Listing 1-18. Using concurrency to speed up the computation of sum

```go
package main

import (
    "fmt"
    "time"
)

const (
    NumbersToSum = 10_000_000
)

func sum(s []float64, c chan<- float64) {
    // A generator that puts data into channel
    sum := 0.0
    for _, v := range s {
        sum += float64(v)
    }
    c <- sum // blocks until c is taken out of the channel
}
```

```go
func plainSum(s []float64) float64 {
    sum := 0.0
    for _, v := range s {
        sum += float64(v)
    }
    return sum
}

func main() {
    s := []float64{}
    for i := 0; i < NumbersToSum; i++ {
        s = append(s, 1.0)
    }

    c := make(chan float64)
    start := time.Now()
    go sum(s[:len(s) / 2], c)
    go sum(s[len(s) / 2 :], c)
    first, second := <-c, <-c // receive from each c
    elapsed := time.Since(start)
    fmt.Printf("first: %f  second: %f elapsed time: %v", first, second,
                elapsed)
    start = time.Now()
    answer := plainSum(s)
    elapsed = time.Since(start)
    fmt.Printf("\nplain sum: %f elapsed time: %v", answer, elapsed)
}
/*
first: 5000000.000000  second: 5000000.000000 elapsed time: 5.864275ms
plain sum: 10000000.000000 elapsed time: 11.601511ms
*/
```

By summing half the numbers in each of two goroutines, a substantial improvement in execution time occurs as is evident in the program output.

The two goroutines perform their computation in a for-loop concurrently and return their values by assigning to the channel variable *c*. In **main**, the assignment of the two sums to **first** and **second** is blocked until both values are assigned to the channel.

Generating Prime Numbers Using Concurrency

Next, we turn to the generation of prime numbers. The classic algorithm for doing this is the Sieve of Eratosthenes. This is an extremely fast nonconcurrent algorithm.

The goal is to find all the prime numbers up to a specified number, say, ten million. A prime number is an integer that is only divisible by 1 or itself. The first several prime numbers are 2, 3, 5, 7, 11, 13, 17, 19, 23. With the exception of 2, all other prime numbers are odd numbers.

Sieve of Eratosthenes Algorithm

The Sieve of Eratosthenes algorithm is presented in the following function:

```go
func SieveOfEratosthenes(n int) []int {
    // Finds all primes up to n
    primes := make([]bool, n+1)
    for i := 2; i < n+1; i++ {
        primes[i] = true
    }
    The sieve logic for removing non-prime indices
    for p := 2; p * p <= n; p++ {
        if primes[p] == true {
            // Update all multiples of p
            for i := p * 2; i <= n; i += p {
                primes[i] = false
            }
        }
    }

    // return all prime numbers <= n
    var primeNumbers []int
    for p := 2; p <= n; p++ {
        if primes[p] == true {
            primeNumbers = append(primeNumbers, p)
        }
    }
    return primeNumbers
}
```

A slice of bool is initialized to contain n + 1 **boolean** values. Each element in the slice is initialized to **true**, indicating that all are initially considered to be primes.

In the for-loop that follows, an index variable **p** is run from 2 up to the square root of **n**. If the value prime[p] is true (indicating that **p** is a prime), all indices that are multiples of *p* are removed from the primes slice. Say, n = 20. When p is 2, the prime slice is set to false at the following indices: 4, 6, 8, 10, 12, 14, 16, 18, 20. When p is 3, the prime slice is set to false at the following indices: 6, 9, 12, 15, 18. When p is 4, the prime slice is set to false at the following indices: 8, 12, 16, 20. Since p = 5 squared exceeds 20, we are done. The sieve has done its work. The indices whose prime values have not been set to false are 2, 3, 5, 7, 11, 13, 17, and 19.

Listing 1-19 presents a program for benchmarking the performance of the sieve.

Listing 1-19. Benchmarking the Sieve of Eratosthenes

```go
// Sieve of Eratosthenes
package main

import (
    "fmt"
    "time"
)

const LargestPrime = 10_000_000

func SieveOfEratosthenes(n int) []int {
    // Finds all primes up to n
    primes := make([]bool, n+1)
    for i := 2; i < n+1; i++ {
        primes[i] = true
    }
    // The Sieve logic
    for p := 2; p*p <= n; p++ {
        if primes[p] == true {
            // Update all multiples of p
            for i := p * 2; i <= n; i += p {
```

```
                primes[i] = false
            }
        }
    }

    // return all prime numbers <= n
    var primeNumbers []int
    for p := 2; p <= n; p++ {
        if primes[p] == true {
            primeNumbers = append(primeNumbers, p)
        }
    }
    return primeNumbers
}

func main() {
    start := time.Now()
    sieve := SieveOfEratosthenes(LargestPrime)
    elapsed := time.Since(start)
    fmt.Println("\nComputation time: ", elapsed)
    fmt.Println("Largest prime: ", sieve[len(sieve)-1])
}
/* Output
Computation time: 28.881792ms
Number of primes = 664579
*/
```

Lest you think that all concurrent solutions are superior, consider the concurrent solution to generating prime numbers in Listing 1-20.

The concurrent solution is so slow compared to the nonconcurrent Sieve of Eratosthenes presented in Listing 1-19 that the constant *LargestPrime* is lowered by two orders of magnitude to 100,000 instead of the 10 million. Even then, the solution is slower.

Listing 1-20. A concurrent daisy chain solution to generating prime numbers

```go
// A concurrent prime sieve
package main

import (
    "fmt"
    "time"
)

const LargestPrime = 100_000

var primes []int
// Send the sequence 3, 5, ... to channel 'ch'.
func Generate(prime1 chan<- int) {
    for i := 3; ; i += 2 {
        prime1 <- i // Send 'i' to channel prime1.
    }
}
// Copy the values from channel 'in' to channel 'out',
// removing those divisible by 'prime'.
func Filter(in <-chan int, out chan<- int, prime int) {
    for {
        i := <-in // Receive value from 'in'.
        if i % prime != 0 {
            out <- i // Send 'i' to 'out'.
        }
    }
}

func main() {
    start := time.Now()
    prime1 := make(chan int) // Create a new channel.
    go Generate(prime1)      // Launch goroutine.
    for {
        prime := <-prime1 // Take prime1 out of channel
        if prime > LargestPrime {
            break
        }
```

```go
        primes = append(primes, prime)
        prime2 := make(chan int)
        go Filter(prime1, prime2, prime)
        prime1 = prime2
    }
    elapsed := time.Since(start)
    fmt.Println("Computation time: ", elapsed)
    fmt.Println("Number of primes = ", len(primes))
}
/* Output
Computation time: 1.462818125s
Number of primes = 9591
*/
```

The use of the remainder operator, **%,** in the ***Filter*** goroutine imposes a significant performance penalty. This goroutine receives information from the ***in*** channel and outputs information to the ***out*** channel as shown by the directional arrows in the signature to the function.

Can we do better by using another concurrent solution?

Segmented Sieve Algorithm

As a stepping stone toward answering this question, we introduce a modification to the Sieve of Eratosthenes algorithm implemented in Listing 1-19. Listing 1-21 presents a segmented sieve algorithm that provides the basis for a concurrent solution to be presented later.

Listing 1-21. Segmented sieve algorithm

package main

```go
import (
    "fmt"
    "math"
    "time"
)
```

const LargestPrime = 10_000_000

```go
var cores int

func SieveOfEratosthenes(n int) []int {
    // Finds all primes up to n
    primes := make([]bool, n+1)
    for i := 2; i < n+1; i++ {
        primes[i] = true
    }

    // The Sieve logic
    for p := 2; p*p <= n; p++ {
        if primes[p] == true {
            // Update all multiples of p
            for i := p * 2; i <= n; i += p {

                primes[i] = false
            }
        }
    }

    // return all prime numbers <= n
    var primeNumbers []int
    for p := 2; p <= n; p++ {
        if primes[p] == true {
            primeNumbers = append(primeNumbers, p)
        }
    }
    return primeNumbers
}

func primesBetween(prime []int, low, high int) []int {
    // Computes the prime numbers between low and high
    // given the initial set of primes from the SieveOfEratosthenes
    limit := high - low
    var result []int
    segment := make([]bool, limit+1)
    for i := 0; i < len(segment); i++ {
        segment[i] = true
    }
```

```go
    // Find the primes in the current segment based on initial primes
    for i := 0; i < len(prime); i++ {
        lowlimit := int(math.Floor(float64(low)/float64(prime[i])) *
                        float64(prime[i]))

        if lowlimit < low {
            lowlimit += prime[i]
        }

        for j := lowlimit; j < high; j += prime[i] {
            segment[j-low] = false
        }
    }

    for i := low; i < high; i++ {
        if segment[i-low] == true {
            result = append(result, i)
        }
    }
    return result
}

func SegmentedSieve(n int) []int {
    // Each segment is of size square root of n
    // Finds all primes up to n
    var primeNumbers []int
    limit := (int)(math.Floor(math.Sqrt(float64(n))))

    prime := SieveOfEratosthenes(limit)
    for i := 0; i < len(prime); i++ {
        primeNumbers = append(primeNumbers, prime[i])
    }

    low := limit
    high := 2 * limit
    if high >= n {
        high = n
    }
```

```go
for {
    if low < n {
        next := primesBetween(prime, low, high)
        // fmt.Printf("\nprimesBetween(%d, %d) = %v", low, high, next)
        for i := 0; i < len(next); i++ {
            primeNumbers = append(primeNumbers, next[i])
        }
        low = low + limit
        high = high + limit
        if high >= n {
            high = n
        }
    } else {
        break
    }
}
    return primeNumbers
}
func main() {
    start := time.Now()
    primeNumbers := SegmentedSieve(LargestPrime)
    elapsed := time.Since(start)
    fmt.Println("\nComputation time: ", elapsed)
    fmt.Println("Number of primes = ", len(primeNumbers))
}
/* Output
Computation time: 50.557584ms
Number of primes = 664579
*/
```

A series of array segments, each of size **square root of n** (where n is the highest number to be considered in the array of primes), are defined. Using the prime numbers in the first such array segment, 0 to sqrt(n), which are computed using the *SieveOfEratosthenes* function, we compute the prime numbers in each succeeding array segment using the *primesBetween* function.

Let us walk through this function when n is 100 and the size of each array segment is 10. Specifically, let us examine the computation of the primes in the segment 10 to 20.

The primes up to 10 are **2**, **3**, **7**.

The variable *low* is 10 and *high* is 20.

An empty slice *result* is defined, and a *segment* slice of bool is created of size **limit + 1**. This *segment* slice is initialized with values of **true**.

In a for-loop ranging from 0 to 3 (the length of **prime**), we define variable *lowlimit* using

int(math.Floor(float64(low)/float64(prime[i]))* float64(prime[i]))

This evaluates to $(10.0 / 2.0) * 2 = 10$.

In another for-loop that ranges from *lowlimit* to *high* in increments of 2, we set segment at indices **10**, **12**, **14**, **16**, **18**, **and 20** to **false**.

We advance the index i from 0 to 1 and compute *lowlimit* as floor(10 / 3) * 3 = 9. Since *lowlimit* is now less than *low*, we set it to 12 using **lowlimit += prime[1]**.

In the loop with index j, we set **segment** slice to false at indices **12**, **15**, and **18**.

Continuing with i set to 2, *lowlimit* is floor(10 / 7) * 7, which equals 7. Since that is less than *low*, we reassign it to 14 (**lowlimt += prime[2]**).

In the j loop, we set the segment slice to false at index **14**.

Finally, we capture the values in the segment slice that are still true: **11**, **13**, **17**, and **19**.

This pattern is the same for each of the segment slices. The number of computations is the same as the original Sieve of Eratosthenes function. But now the array slice is much smaller (size 10 instead of size 100).

As you can see from the program output, the segmented sieve solution is 1.75 times slower (50.557584ms) compared to the original Sieve of Eratosthenes solution (28.881792ms). This is due to the overhead of defining the ten *segment* slices and the overhead of the function calls to these slices, not needed in the original sieve computation.

The stage has been set now for a concurrent solution that leverages off the segmented sieve.

Concurrent Sieve Solution

This concurrent solution is presented in Listing 1-22.

Listing 1-22. Concurrent segmented sieve

```go
package main

import (
    "fmt"
    "math"
    "runtime"
    "sync"
    "time"
)

const LargestPrime = 10_000_000

var cores int
var primeNumbers []int
var m sync.Mutex
var wg sync.WaitGroup

func SieveOfEratosthenes(n int) []int {
    // Finds all primes up to n
    primes := make([]bool, n+1)
    for i := 2; i < n+1; i++ {
        primes[i] = true
    }

    // The Sieve logic
    for p := 2; p*p <= n; p++ {
        if primes[p] == true {
            // Update all multiples of p
            for i := p * 2; i <= n; i += p {

                primes[i] = false
            }
        }
    }
```

```go
    // return all prime numbers <= n
    var primeNumbers []int
    for p := 2; p <= n; p++ {
        if primes[p] == true {
            primeNumbers = append(primeNumbers, p)
        }
    }
    return primeNumbers
}

func primesBetween(prime []int, low, high int) {
    // Computes the prime numbers between low and high
    // given the initial set of primes from the SieveOfEratosthenes
    defer wg.Done()
    limit := high - low
    segment := make([]bool, limit+1)
    for i := 0; i < len(segment); i++ {
        segment[i] = true
    }
    // Find the primes in the current segment based on initial primes
    for i := 0; i < len(prime); i++ {
        lowlimit := int(math.Floor(float64(low)/float64(prime[i])) *
                        float64(prime[i]))
        if lowlimit < low {
            lowlimit += prime[i]
        }

        for j := lowlimit; j < high; j += prime[i] {
            segment[j-low] = false
        }

        // Each number in [low to high] is mapped to [0, high - low]
        for j := lowlimit; j < high; j += prime[i] {
            segment[j-low] = false
        }
    }

    m.Lock()
```

```go
        for i := low; i < high; i++ {
            if segment[i-low] == true {
                primeNumbers = append(primeNumbers, i)
            }
        }
    }
    m.Unlock()
}

func SegmentedSieve(n int) {
    limit := int(math.Floor(float64(n) / float64(cores)))
    prime := SieveOfEratosthenes(limit)
    for i := 0; i < len(prime); i++ {
        primeNumbers = append(primeNumbers, prime[i])
    }

    for low := limit; low < n; low += limit {
        high := low + limit
        if high >= n {
            high = n
        }
        wg.Add(1)
        go primesBetween(prime, low, high)
    }
    wg.Wait()
}

func main() {
    cores = runtime.NumCPU()
    start := time.Now()
    SegmentedSieve(LargestPrime)
    elapsed := time.Since(start)
    fmt.Println("\nComputation time for concurrrent: ", elapsed)
    fmt.Println("Number of primes = ", len(primeNumbers))
}

/* Output
Computation time for concurrrent: 19.783666ms
Number of primes = 664579
*/
```

The concurrency is achieved in function SegmentedSieve, which launches a series of goroutines, *primesBetween*, and uses a **WaitGroup** to block the completion of *SegmentedSieve* until all the goroutines have completed their work.

To prevent a race condition from occurring, a mutex, *m*, is used at the end of each goroutine to guarantee that the assignment to the globally shared *primeNumbers* slice is controlled using an *m.Lock()* at the beginning of the assignment loop and an *m.UnLock()* at the end of this assignment loop.

The time required to obtain prime numbers up to the number 10 million is 19.78366ms, which is smaller than the Sieve of Eratosthenes computation, which requires 28.881792ms. The segment size is computed by choosing a number of cores given by *runtime.NumCPU()*. In principle, this should allow a computation that utilizes each of the computer cores approximating parallel processing. The use of the mutex to protect against a race condition compromises the efficiency of the concurrent solution but is essential using the approach taken to avoid a race condition.

The sequence of primes that are generated using the concurrent segmented sieve is not in order but is complete. This is because the goroutines run asynchronously in random order.

1.5 Summary

In this chapter, we introduced and illustrated the use of generic types. We demonstrated that using generic types can greatly simplify application development by avoiding duplication of code each time we change an underlying type used by the code. We set the stage for using generic types in the data structures and algorithms to be presented throughout this book.

We also demonstrated the potential benefits of using concurrency. We showed that the use of concurrency does not automatically guarantee improved performance. We looked at the use of goroutines and channels as a vehicle of communication between goroutines. We introduced the mutex as a construct for avoiding race conditions and the WaitGroup as a construct for assuring some synchronization between goroutines.

In the next chapter, we enter the world of algorithm design. We discuss methods for characterizing algorithm efficiency. We look at some classic sorting algorithms and the use of concurrency to attain faster sorting.

## CHAPTER 2

Algorithm Efficiency: Sorting and Searching

The previous chapter introduced generics and reviewed concurrency. We utilize both going forward in this chapter and the rest of the book.

The principal goal of this book is providing techniques based on data structures and algorithms for making programs run faster or in less space (more efficiently). The first question we address in this chapter is how we describe the efficiency of an algorithm. We then examine sorting and searching algorithms and examine their efficiency.

2.1 Describing the Speed Efficiency of an Algorithm

The normal practice in determining the efficiency of an algorithm is to estimate its performance as a function of problem size, asymptotically. That is, we are concerned with the speed of computation as the size of the problem becomes significant. In this section, we'll introduce Big O notation and its application in algorithm design.

Working with Big O

Big O notation characterizes how the execution time of an algorithm grows as a function of problem size as the problem size becomes large. It is based on an analysis of how many basic operations such as assignment, swapping, and accessing a value are required to perform the task. Because of the requirement that the problem size must become large, big O is an asymptotic performance indicator.

For example, suppose we were able to characterize the runtime of some algorithm as a function of problem size n as follows: execution-time(n) = $12n^2 + 117n + 25$.

© Richard Wiener 2022
R. Wiener, *Generic Data Structures and Algorithms in Go*, https://doi.org/10.1007/978-1-4842-8191-8_2

When n is large, the first term in the preceding expression dominates. We ignore the constant 12 and focus on the n^2. This leads us to characterize the algorithm as **O(n^2)**. For large n, if we were to double the value of n, the execution time would quadruple. This would not hold if n were small.

Big O provides an asymptotic upper bound. The actual performance for large n (problem size) is bounded by the function inside the O notation. So **O(n)** implies that for large n, the algorithm's execution time is bounded by n or is linear with respect to n.

Algorithms with **O(2^n)**, exponential complexity, are intractable. Likewise, algorithms of O(n!) are intractable. As the size of the problem, n, becomes large, the computation time becomes too great to provide any reasonable completion. We examine and tackle computationally intractable problems later in this book.

We return to computationally tractable problems now and consider the following example. We wish to determine whether an array slice of floating-point numbers is sorted from smallest to largest.

Determining Whether a Slice of Numbers Is Sorted

One approach to solving this problem is to use the **Sort** function in package **sort** and then compare the resultant array slice with the array slice we wish to test. This is an approach that I have seen many less experienced programmers take.

A function that performs this is given as follows:

```
func isSorted1(data []float64) bool {
    var data1 []float64
    data1 = make([]float64, len(data)) // Creates a slice of len(data)
    copy(data1, data) // Copies data into data1
    sort.Float64s(data1)

    // Compare data and data1
    for i := 0; i < size; i++ {
        if data[i] != data1[i] {
            return false
        }
    }
    return true
}
```

We allocate storage for the *data1* slice and copy the input *data* into *data1*. We sort *data1* using *sort.Float64s(data1)*. Finally, we compare each element of *data1* with *data* and return false if there is a mismatch.

It is known that the asymptotic complexity of the **Sort** algorithm is **O(nlog$_2$n)**.

Now consider as an alternative approach function *isSorted2*, given as follows:

```
func isSorted2(data []float64) bool {
    for i := 1; i < len(data); i++ {
        if data[i] < data[i - 1] {
            return false
        }
    }
    return true
}
```

This function compares all consecutive pairs of *data*. If an instance of **data[i]** is less than **data[i – 1]**, the function immediately returns false. If no returns of false occur, the function returns true.

This function has an asymptotic complexity of **O(n)**. For large n, *isSorted2* should be much faster than *isSorted1*.

Listing 2-1 does a benchmark comparison between the two **isSorted** functions.

Listing 2-1. Comparing two isSorted functions

```
package main

import (
    "fmt"
    "math/rand"
    "time"
    "sort"
)

const size = 100_000_000
var data []float64

func isSorted1(data []float64) bool {
    var data1 []float64
    data1 = make([]float64, len(data))
```

```go
    copy(data1, data) // Copies data into data1
    sort.Float64s(data1)

    // Compare data and data1
    for i := 0; i < size; i++ {
        if data[i] != data1[i] {
            return false
        }
    }
    return true
}

func isSorted2(data []float64) bool {
    for i := 1; i < len(data); i++ {
        if data[i] < data[i - 1] {
            return false
        }
    }
    return true
}

func main() {
    data = make([]float64, size)
    for i := 0; i < size; i++ {
        data[i] = 100.0 * rand.Float64()
    }

    start := time.Now()
    result := isSorted1(data)
    elapsed := time.Since(start)
    fmt.Println("Sorted: ", result)
    fmt.Println("elapsed  using sorted1:", elapsed)

    data2 := make([]float64, size)
    for i := 0; i < size; i++ {
        data2[i] = float64(2 * i)
    }
```

```
    start = time.Now()
    result = isSorted1(data2)
    elapsed = time.Since(start)
    fmt.Println("Sorted: ", result)
    fmt.Println("elapsed using sorted1:", elapsed)

    start = time.Now()
    result = isSorted2(data)
    elapsed = time.Since(start)
    fmt.Println("\nSorted: ", result)
    fmt.Println("elapsed using sorted2", elapsed)

    start = time.Now()
    result = isSorted2(data2)
    elapsed = time.Since(start)
    fmt.Println("Sorted: ", result)
    fmt.Println("elapsed using sorted2:", elapsed)
}
/* Output
Sorted:   false
elapsed  using sorted1: 20.554518978s
Sorted:   true
elapsed using sorted1: 7.328819941s

Sorted:   false
elapsed using sorted2 291ns
Sorted:   true
elapsed using sorted2: 76.644396ms
*/
```

Each function is invoked with an array, *data1*, of random **float64** values. Next, each function is invoked with an array that is already sorted, *data2*.

As evident in the output, *isSorted2* is over 100 times faster than *isSorted1*, confirming the analysis with big O.

Using Concurrency

Can we do better by using concurrency? Consider function *isSorted3* as follows:

```go
func isSegmentSorted(data []float64, a, b int, ch chan<- bool) {
    // Generates boolean value put into ch
    for i := a + 1; i < b; i++ {
        if data[i] < data[i - 1] {
            ch <- false
        }
    }
    ch <- true
}

func isSorted3(data []float64) bool {
    ch := make(chan bool)
    numSegments := runtime.NumCPU()
    segmentSize := int(float64(len(data)) / float64(numSegments))
    // Launch numSegments goroutines
    for index := 0; index < numSegments; index++ {
        go isSegmentSorted(data, index * segmentSize,
                index * segmentSize + segmentSize, ch)
    }
    num := 0 // completed goroutines
    for {
        select {
        case value := <- ch:  // Blocks until a goroutine puts a bool into the
                              // channel
            if value == false {
                return false
            }
            num += 1
            if num == numSegments { // All goroutiines have completed
                return true
            }
```

```
        }
    }
    return true
}
```

In function *isSorted3*, we subdivide the data into *numberSegments* given by the number of CPUs on the computer. In a for-loop, we launch *numSegments* goroutines, passing the starting and ending indices, *a* and *b*, along with a channel variable, *ch*, of type **bool**.

Each goroutine uses the same logic as in function *isSorted2* over a much smaller interval and concurrent with the other goroutines. Each goroutine assigns its result to the channel variable *ch*.

In a for-loop in function *isSorted3*, a **select** statement reads a Boolean value from the channel *ch* and blocks program execution until another goroutine has completed its work. If a value of **false** is received, *isSorted3* immediately returns **false**. If not, the value of *num*, which counts the goroutines that have completed their work, is incremented by one. If *num* reaches *numberSegments*, *isSorted3* returns true since all segments have reported true.

In Listing 2-2, we extend the benchmark test to include the concurrent *isSorted3*.

Listing 2-2. Concurrent implementation and timing of isSorted

```
package main

import (
    "fmt"
    "math/rand"
    "time"
    "sort"
    "runtime"
)

const size = 1_000_000_000

var data []float64

// Snip
```

```go
func isSorted3(data []float64) bool {
    ch := make(chan bool)
    numSegments := runtime.NumCPU()
    segmentSize := int(float64(len(data)) / float64(numSegments))
    // Launch numSegments goroutines
    for index := 0; index < numSegments; index++ {
        go isSegmentSorted(data, index * segmentSize,
                    index * segmentSize + segmentSize, ch)
    }
    num := 0 // Completed goroutines
    for {
        select {
        case value := <- ch:
            if value == false {
                return false
            }
            num += 1
            if num == numSegments { // All goroutiines have completed
                return true
            }
        }
    }
    return true
}

func main() {

    data = make([]float64, size)

    for i := 0; i < size; i++ {
        data[i] = 100.0 * rand.Float64()
    }

    data2 := make([]float64, size)
    // Create a sorted sequence of numbers
    for i := 0; i < size; i++ {
        data2[i] = float64(2 * i)
    }
```

```go
    start := time.Now()
    result := isSorted2(data)
    elapsed := time.Since(start)
    fmt.Println("\nSorted: ", result)
    fmt.Println("elapsed using sorted2", elapsed)
    start = time.Now()
    result = isSorted2(data2)
    elapsed = time.Since(start)
    fmt.Println("Sorted: ", result)
    fmt.Println("elapsed using sorted2:", elapsed)

    start = time.Now()
    result = isSorted3(data)
    elapsed = time.Since(start)
    fmt.Println("\nSorted: ", result)
    fmt.Println("elapsed using concurrent sorted3", elapsed)
    start = time.Now()
    result = isSorted3(data2)
    elapsed = time.Since(start)
    fmt.Println("Sorted: ", result)
    fmt.Println("elapsed using concurrent sorted3:", elapsed)
}
/* Output
Sorted:  false
elapsed using sorted2 594ns
Sorted:  true
elapsed using sorted2: 845.586082ms

Sorted:  false
elapsed using concurrent sorted3 61.863µs
Sorted:  true
elapsed using concurrent sorted3: 132.375156ms
*/
```

The size of the array to test for *isSorted* has been increased to a billion floating-point numbers.

The results are dramatic. The concurrent *isSorted* solution is over six times faster than the noncurrent solution. Both solutions are of **O(n)**. Improving performance by a constant factor does not change the big O complexity of the algorithm.

In the next section, we present several classic sorting algorithms and their complexity.

2.2 Sorting Algorithms

Sorting collections of data such as a slice in Go has always been a fundamental part of learning computer science. In this section, we look at two well-known sorting algorithms and examine their complexity using a big O analysis.

Bubblesort Algorithm

Listing 2-3 implements a generic bubblesort algorithm assuming an ordered slice of data (base type where each element can be compared with respect to greater than or less than).

Listing 2-3. Generic bubble sort

```go
package main

import(
    "fmt"
)

type Ordered interface {
    ~float64 | ~int | ~string
}

func bubblesort[T Ordered](data []T) {
    n := len(data)
    for i:= 0; i < n - 1; i++ {
        for j:= 0; j < n - 1 - i; j++ {
            if data[j] > data[j + 1] {
```

```
          data[j], data[j + 1] = data[j + 1], data[j]
        }
      }
    }
}

func main() {
    numbers := []float64{3.5, -2.4, 12.8, 9.1}
    names := []string{"Zachary", "John", "Moe", "Jim", "Robert"}
    bubblesort[float64](numbers)
    fmt.Println(numbers)
    bubblesort[string](names)
    fmt.Println(names)
}
/* Output
[-2.4 3.5 9.1 12.8]
[Jim John Moe Robert Zachary]
*/
```

The type ***Ordered*** can have many more basic types included. The tilde symbol in front of each of the basic types means that any user-defined type that uses the given base type is considered ***Ordered***.

Bubblesort has earned its popularity in CS 1 courses because of its relative simplicity. Elements are compared sequentially and interchanged if out of order. On each iteration of the outer loop, the largest value "bubbles" to the rightmost position in the slice. This position is not considered during the next iteration of the inner loop because of

j < n - 1 - i.

The nested for-loops, shown in boldface, make this an **O(n²)** algorithm. In general, k nested loops produce an algorithm of **O(n^k)**.

Bubblesort is most efficient when the slice being sorted is already sorted and slowest when the slice is in reverse order.

Next, we examine one of the most widely used sorting algorithms, the classic **quicksort**.

65

Quicksort Algorithm

As the name implies, this algorithm is reputed to perform very fast sorts.

Listing 2-4 shows the implementation of a generic quicksort.

Listing 2-4. Generic quicksort

```go
package main

import(
    "fmt"
)

type Ordered interface {
    ~float64 | ~int | ~string
}

func quicksort[T Ordered](data []T, low, high int) {
    if low < high {
        var pivot = partition(data, low, high)
        quicksort(data, low, pivot)
        quicksort(data, pivot + 1, high)
    }
}

func partition[T Ordered](data []T, low, high int) int {

    // Pick a lowest bound element as a pivot value
    var pivot = data[low]

    var i = low
    var j = high

    for i < j {
        for data[i] <= pivot && i < high {
            i++;
        }

        for data[j] > pivot && j > low {
            j--
        }
```

```go
        if i < j {
            data[i], data[j] = data[j], data[i]
        }
    }
    data[low] = data[j]
    data[j] = pivot
    return j
}

func main() {
    numbers := []float64{3.5, -2.4, 12.8, 9.1}
    names := []string{"Zachary", "John", "Moe", "Jim", "Robert"}
    quicksort[float64](numbers, 0, len(numbers) - 1)
    fmt.Println(numbers)
    quicksort[string](names, 0, len(names) - 1)
    fmt.Println(names)
}
/* Output
[-2.4 3.5 9.1 12.8]
[Jim John Moe Robert Zachary]
*/
```

The **quicksort** algorithm is an example of a divide-and-conquer algorithm. We partition the original slice into two smaller slices and sort each of these by continuing to partition each into two more until eventually we get slices of two elements that we compare with each other.

If the original slice has n elements, we can perform the divide-and-conquer $\log_2 n$ times (the number of times we can divide n by 2 to get down to two elements).

This assumes that we partition the slices by cutting them in half each time. A close examination of the *partition* function reveals that this is not always the case.

The partition function uses its leftmost element as the pivot element. It then moves data around the slice to ensure that elements to the left of the pivot element are smaller and elements to the right of the pivot are larger.

Big O Analysis

We "walk" through an example to illustrate, in detail, how partition does its work.

Suppose our array slice that we wish to partition is

```
[6, 2, 3, 9, 8, 17, 4]
```

We choose the pivot element to be 6 (the leftmost element).

We increment the index i until we find an element larger than the pivot element 6. That is element 9 in position 3.

Now starting with index j in position 6 (the rightmost position), we decrement j until we find an element whose value is less than the pivot element. That is element 4 in position 6. We interchange the elements in positions 3 and 6 producing

```
[6, 2, 3, 4, 8, 17, 9]
```

Starting at index 3, we again increment i until we find an element greater than 6, which is 8 in position 4. We decrement j (in position 6) until we find an element less than the pivot element 6. That is element 4 in position 3. We don't interchange the elements in positions 3 and 4 since i is not less than j.

Since i is no longer less than j, we exit the outer for-loop.

We perform the final interchange of the pivot element with position j to get

```
[4, 2, 3, 6, 8, 17, 9]
```

All the elements to the left of the pivot element 6 are less than 6, and those to the right of 6 are greater than 6.

Worst Case for Quicksort

If the original array were sorted, so the pivot element was the smallest, this would be worst case. We would have to interchange n − 1 elements on the first pass, n − 2 elements on the second pass, etc., giving us an $O(n^2)$ algorithm. This is one of the ironies of quicksort. The closer the data is to initially sorted, the worse quicksort performs.

A useful filter to impose in front of quicksort would be to test the input to see whether it is already sorted. If so, bail out and don't perform any sorting.

Since the partition function is $O(n)$, the quicksort algorithm is $O(n\log_2 n)$ when the data being sorted is not already sorted or close to sorted.

Comparing Bubblesort to Quicksort

In Listing 2-5, we compare **bubblesort** with **quicksort** using sine wave data for our array slice.

Listing 2-5. Comparing bubblesort with quicksort

```
package main

import(
    "fmt"
    "math"
    "time"
)

const size = 100_000

type Ordered interface {
    ~float64 | ~int | ~string
}

// Snip - See Listings 2.3 and 2.4

func main() {
    data := make([]float64, size)
    for i := 0; i < size; i++ {
        data[i] = math.Sin(float64(i * i))
    }
    start := time.Now()
    quicksort[float64](data, 0, len(data) - 1)
    elapsed := time.Since(start)
    fmt.Println("Elapsed sort time for sine wave using quicksort: ",
    elapsed)

    data = make([]float64, size)
    for i := 0; i < size; i++ {
        data[i] = math.Sin(float64(i * i))
    }
```

```
    start = time.Now()
    bubblesort[float64](data)
    elapsed = time.Since(start)
    fmt.Println("Elapsed sort time for sine wave using bubblesort: ",
    elapsed)
}
/*Output
lapsed sort time for sine wave using quicksort:  7.808522ms
Elapsed sort time for sine wave using bubblesort:  12.26859692s
*/
```

The $O(nlog_2 n)$ quicksort performs almost 1600 times faster than the $O(n^2)$ bubblesort.

Concurrent Quicksort

Can we improve the performance of **quicksort** using concurrency?

As a stepping stone toward a concurrent solution, we consider another $O(n^2)$ sorting algorithm, *InsertSort*, implemented as follows:

```
func InsertSort[T Ordered](data[] T) {
    i := 1
    for i < len(data) {
        h := data[i]
        j := i - 1
        for j >= 0 && h < data[j] {
            data[j + 1] = data[j]
            j -= 1
        }
        data[j + 1] = h
        i += 1
    }
}
```

Let us "walk" through a simple example to see how this sorting algorithm works.
Suppose the slice, *data*, that we wish to sort is **[5, 1, 12, 9]**.
The variable **i** is initialized to 1.

In the for-loop, **h** is set to **data[1]**, which is 1. The variable **j** is set to 0. In a nested for-loop, **data[j + 1]** is set to **data[0]**, so **data[1]** is set to 5. The inner loop terminates. Next, **data[0]** is set to 1. The slice is now **[1, 5, 12, 9]**.

After incrementing **i**, we execute the outer loop again. The variable **h** is set to 12, and **j** is set to 1. The inner loop terminates since 12 is not less than either 1 or 5. We increment **i** to 3. Variable **h** is set to 9. In the inner loop, since 9 is less than 12, the slice becomes **[1, 5, 9, 12]**, and we are done.

Because of the two nested loops, the **InsertSort** is **O(n²)** for large n.

For each iteration of the outer loop, the next element, not yet considered, is inserted to the left of the first element that is larger than it.

In Listing 2-6, we consider a concurrent implementation of quicksort of 50 million numbers.

Listing 2-6. Concurrent implementation of quicksort

```go
package main
import (
    "fmt"
    "time"
    "math/rand"
    "sync"
)

const size = 50_000_000
const threshold = 5000

type Ordered interface {
    ~float64 | ~int | ~string
}

func InsertSort[T Ordered](data[] T) {
    i := 1
    for i < len(data) {
        h := data[i]
        j := i - 1
        for j >= 0 && h < data[j] {
            data[j + 1] = data[j]
            j -= 1
```

```
        }
        data[j + 1] = h
        i += 1
    }
}

func Partition[T Ordered](data[] T) int {
    data[len(data) / 2], data[0] = data[0], data[len(data) / 2]
    pivot := data[0]
    mid := 0
    i := 1
    for i < len(data) {
        if data[i] < pivot {
            mid += 1
            data[i], data[mid] = data[mid], data[i]
        }
        i += 1
    }
    data[0], data[mid] = data[mid], data[0]
    return mid
}

func IsSorted[T Ordered](data[] T) bool {
    for i := 1; i < len(data); i++ {
        if data[i] < data[i - 1] {
            return false
        }
    }
    return true
}

func ConcurrentQuicksort[T Ordered](data[] T, wg *sync.WaitGroup) {
    for len(data) >= 30 {
        mid := Partition(data)
        var portion[] T
        if mid < len(data) / 2 {
            portion = data[:mid]
```

```go
            data = data[mid + 1:]
        } else {
            portion = data[mid + 1:]
            data = data[:mid]
        }
        if (len(portion) > threshold) {
            wg.Add(1)
            go func(data[] T) {
                defer wg.Done()
                ConcurrentQuicksort(data, wg)
            }(portion)
        } else {
            ConcurrentQuicksort(portion, wg)
        }
    }
    InsertSort(data)
}

func QSort[T Ordered](data[] T) {
    var wg sync.WaitGroup
    ConcurrentQuicksort(data, &wg)
    wg.Wait()
}

func partition[T Ordered](data []T, low, high int) int {
    var pivot = data[low]
    var i = low
    var j = high
    for i < j {
        for data[i] <= pivot && i < high {
            i++;
        }
        for data[j] > pivot && j > low {
            j--
        }
        if i < j {
```

```go
            data[i], data[j] = data[j], data[i]
        }
    }
    data[low] = data[j]
    data[j] = pivot
    return j
}

func quicksort[T Ordered](data []T, low, high int) {
    if low < high {
        var pivot = partition(data, low, high)
        quicksort(data, low, pivot)
        quicksort(data, pivot + 1, high)
    }
}

func main() {
    data := make([]float64, size)
    for i := 0; i < size; i++ {
        data[i] = 100.0 * rand.Float64()
    }
    data2 := make([]float64, size)
    copy(data2, data)

    start := time.Now()
    QSort[float64](data)
    elapsed := time.Since(start)
    fmt.Println("Elapsed time for concurrent quicksort = ", elapsed)
    fmt.Println("Is sorted: ", IsSorted(data))

    start = time.Now()
    quicksort(data2, 0, len(data2) - 1)
    elapsed = time.Since(start)
    fmt.Println("Elapsed time for regular quicksort = ", elapsed)
    fmt.Println("Is sorted: ", IsSorted(data2))
}
/* Output
```

```
Elapsed time for concurrent quicksort = 710.431619ms
Is sorted:  true
Elapsed time for regular quicksort = 5.382400384s
Is sorted:  true
*/
```

The results are again dramatic. In sorting 50 million numbers and comparing the sort time of regular quicksort with that of concurrent quicksort, we find the regular quicksort is about 7.6 times slower than the concurrent quicksort.

When the length of the slice is less than 30, we use *InsertSort* to complete the sorting of the slice. When the length of the slice is less than the threshold of 5000, we use ordinary quicksort to complete the sorting. The constants 30 and 5000 are determined empirically. The motivation is to prevent the overhead associated with many goroutines deployed to sort small-sized slices.

It is noted that the performance of *InsertSort* on a slice whose size is less than 30 is not governed by $O(n^2)$, which is an asymptotic bound for large **n**.

Mergesort Algorithm

The next sorting algorithm we examine is the classic **mergesort** algorithm. It is a divide-and-conquer algorithm, just like quicksort. We replace the original slice with two slices, each of size a half of the original slice size. Each of these half-slices is further divided into quarter slices, and this pattern continues until we get slices of size 1. Using recursion, we weave the slices together by merging them as shown in Listing 2-7.

Listing 2-7. Mergesort algorithm

```
package main

import (
    "fmt"
    "math/rand"
    "time"
)
```

```
const size = 50_000_000

type Ordered interface {
    ~float64 | ~int | ~string
}

func IsSorted[T Ordered](data[] T) bool {
    for i := 1; i < len(data); i++ {
        if data[i] < data[i - 1] {
            return false
        }
    }
    return true
}

func InsertSort[T Ordered](data[] T) {
    i := 1
    for i < len(data) {
        h := data[i]
        j := i - 1
        for j >= 0 && h < data[j] {
            data[j + 1] = data[j]
            j -= 1
        }
        data[j + 1] = h
        i += 1
    }
}

func Merge[T Ordered](left, right []T) []T {
    result := make([]T, len(left) + len(right))
    i, j, k := 0, 0, 0

    for i < len(left) && j < len(right) {
        if left[i] < right[j] {
            result[k] = left[i]
            i++
        } else {
```

```
            result[k] = right[j]
            j++
        }
        k++
    }
    for i < len(left) {
        result[k] = left[i]
        i++
        k++
    }
    for j < len(right) {
        result[k] = right[j]
        j++
        k++
    }
    return result
}

func MergeSort[T Ordered](data []T) []T {
    if len(data) > 100 {
        middle := len(data) / 2
        left := data[:middle]
        right := data[middle:]
        data = Merge(MergeSort(right), MergeSort(left))
    } else {
        InsertSort(data)
    }
    return data
}

func main() {
    data := make([]float64, size)
    for i := 0; i < size; i++ {
        data[i] = 100.0 * rand.Float64()
    }
    /*
```

```go
    data2 := make([]float64, size)
    copy(data2, data)
    */

    start := time.Now()
    result := MergeSort[float64](data)
    elapsed := time.Since(start)
    fmt.Println("Elapsed time for MergeSort = ", elapsed)
    fmt.Println("Is sorted: ", IsSorted(result))
}
/* Output
Elapsed time for MergeSort = 6.18063849s
Is sorted: true
*/
```

This algorithm is simpler to understand than quicksort. Function *Merge* constructs a new slice, *result*, by merging elements from the two input slices, *left* and *right*, so that *result* is sorted.

The recursive *MergeSort* function partitions the input array into *left* and *right* and calls *Merge* on the results of recursively invoking *MergeSort*.

It is noted that *MergeSort* does not sort in place as **quicksort** does. This requires extra memory allocation compared to **quicksort**.

Since Merge is **O(n)** and there are **$\log_2 n$** recursive calls in *MergeSort*, the complexity of *MergeSort* is **O(n$\log_2$n)**.

Concurrent Mergesort

Can we improve the performance of *MergeSort* with concurrency? Yes!

Listing 2-8 shows a concurrent implementation of *MergeSort*.

Listing 2-8. Concurrent implementation of MergeSort

package main

```go
import (
    "fmt"
```

```go
        "time"
        "math/rand"
        "sync"
)

const size = 50_000_000
const max = 5000

type Ordered interface {
    ~float64 | ~int | ~string
}

func IsSorted[T Ordered](data[] T) bool {
    for i := 1; i < len(data); i++ {
        if data[i] < data[i - 1] {
            return false
        }
    }
    return true
}

func InsertSort[T Ordered](data[] T) {
    i := 1
    for i < len(data) {
        h := data[i]
        j := i - 1
        for j >= 0 && h < data[j] {
            data[j + 1] = data[j]
            j -= 1
        }
        data[j + 1] = h
        i += 1
    }
}

func Merge[T Ordered](left, right []T) []T {
    result := make([]T, len(left) + len(right))
    i, j, k := 0, 0, 0
```

```
    for i < len(left) && j < len(right) {
        if left[i] < right[j] {
            result[k] = left[i]
            i++
        } else {
            result[k] = right[j]
            j++
        }
        k++
    }
    for i < len(left) {
        result[k] = left[i]
        i++
        k++
    }
    for j < len(right) {
        result[k] = right[j]
        j++
        k++
    }
    return result
}

func MergeSort[T Ordered](data []T) []T {
    if len(data) > 100 {
        middle := len(data) / 2
        left := data[:middle]
        right := data[middle:]
        data = Merge(MergeSort(right), MergeSort(left))
    } else {
        InsertSort(data)
    }
    return data
}
```

```go
func ConcurrentMergeSort[T Ordered](data []T) []T {
    if len(data) > 1 {
        if len(data) <= max {
            return MergeSort(data)
        } else { // Concurrent
            middle := len(data) / 2
            left := data[:middle]
            right := data[middle:]
            var wg sync.WaitGroup
            wg.Add(2)
            var data1, data2 []T
            go func() {
                defer wg.Done()
                data1 = ConcurrentMergeSort(left)
            }()

            go func() {
                defer wg.Done()
                data2 = ConcurrentMergeSort(right)
            }()

            wg.Wait()
            return Merge(data1, data2)
        }
    }
    return nil
}

func main() {
    data := make([]float64, size)
    for i := 0; i < size; i++ {
        data[i] = 100.0 * rand.Float64()
    }
    start := time.Now()
    result := ConcurrentMergeSort(data)
    elapsed := time.Since(start)
    fmt.Println("Elapsed time for concurrent mergesort = ", elapsed)
```

```
    fmt.Println("Sorted: ", IsSorted(result))
}
/* Output
Elapsed time for concurrent mergesort = 1.275120179s
Sorted:   true
*/
```

The two goroutines, shown in boldface, perform **MergeSort** concurrently. The *wg. Wait()* forces the *Merge* of the two results to wait for both goroutines to finish.

To avoid the overhead of spawning goroutines for small-sized *data*, ordinary sequential *MergeSort* is used when *data* has a size less than *max* (5000 in this case).

The performance of *ConcurrentMergeSort* on a random slice of 50 million floating-point numbers is slightly slower than *ConcurrentQuickSort* but faster than the sequential version.

Conclusions

Quicksort has an average complexity of **O(nlog₂n)** and sorts in place (no need for extra storage). If the input data is already sorted or close to sorted, the complexity falls to **O(n²)**. The concurrent quicksort is extremely fast.

MergeSort has an average complexity of **O(nlog₂n)**. It does not sort in place, so there is a need for extra storage. Generally, it is slower than quicksort. If the input data is sorted or close to sorted, mergesort is very fast. The concurrent mergesort is extremely fast but slower than the concurrent quicksort.

In the next section, we examine the issue of searching array slices.

2.3 Searching Array Slices

In this section, we restrict our attention to searching array slices efficiently.

Searching a data structure for the presence of stored information is one of the important operations we perform and is often the reason we create the data structure. As we introduce data structures in later sections of the book, we examine methods for efficiently searching for information stored in the data structure.

Linear Searches

The simplest search algorithm for searching a slice is a linear search. We iterate through all the elements of the slice sequentially until we find a matchup or complete searching all the elements of the slice.

Listing 2-9 presents the linear search in a slice.

Listing 2-9. Linear search of a slice

```go
package main

import (
    "fmt"
    "time"
    "math/rand"
)

const size = 100_000_000

type Ordered interface {
    ~float64 | ~int | ~string
}

func linearSearch[T Ordered](slice []T, target T) bool {
    // Return true if T is in the slice
    for i := 0; i < len(slice); i++ {
        if slice[i] == target {
            return true
        }
    }
    return false
}
func main() {
    data := make([]float64, size)
    for i := 0; i < size; i++ {
        data[i] = 100.0 * rand.Float64()
    }
    start := time.Now()
```

```
    result := linearSearch[float64](data, 54.0)
    elapsed := time.Since(start)
    fmt.Println("Time to search slice of 100_000_000 floats using
    linearSearch = ", elapsed)
    fmt.Println("Result of search is ", result)

    start = time.Now()
    result = linearSearch[float64](data, data[size / 2])
    elapsed = time.Since(start)
    fmt.Println("Time to search slice of 100_000_000 floats using
    linearSearch = ", elapsed)
    fmt.Println("Result of search is ", result)
}
/* Output
Time to search slice of 100_000_000 floats using linearSearch = 54.464458ms
Result of search is false
Time to search slice of 100_000_000 floats using linearSearch = 17.981833ms
Result of search is true
*/
```

The preceding benchmark was done on a MacBook Pro with M1 Max processor and 32G of RAM.

Concurrent Searches

In Listing 2-10, we show the details of a concurrent search of an array slice.

Listing 2-10. Concurrent search of a slice

```
package main

import (
    "fmt"
    "time"
    "math/rand"
    "runtime"
)
```

```go
type Ordered interface {
    ~float64 | ~int | ~string
}

const size = 100_000_000

func searchSegment[T Ordered](slice []T, target T, a, b int, ch
chan<- bool) {
    // Generates boolean value put into ch
    for i := a; i < b; i++ {
        if slice[i] == target {
            ch <- true
        }
    }
    ch <- false
}

func concurrentSearch[T Ordered](data []T, target T) bool {
    ch := make(chan bool)
    numSegments := runtime.NumCPU()
    segmentSize := int(float64(len(data)) / float64(numSegments))
    // Launch numSegments goroutines
    for index := 0; index < numSegments; index++ {
        go searchSegment(data, target, index * segmentSize, index *
                            segmentSize + segmentSize, ch)
    }
    num := 0 // Completed goroutines
    for {
        select {
        case value := <- ch:   // Blocks until a goroutine puts a bool into the
                            //channel
            if value == true {
                return true
            }
            num += 1
```

```go
            if num == numSegments { // All goroutiines have completed
                return false
            }
        }
    }
    return false
}

func main() {
    data := make([]float64, size)
    for i := 0; i < size; i++ {
        data[i] = 100.0 * rand.Float64()
    }
    start := time.Now()
    result := concurrentSearch[float64](data, 54.0) // Should return false
    elapsed := time.Since(start)
    fmt.Println("Time to search slice of 100_000_000 floats using
                    concurrentSearch = ", elapsed)
    fmt.Println("Result of search is ", result)

    start = time.Now()
    result = concurrentSearch[float64](data, data[size / 2]) // true
    elapsed = time.Since(start)
    fmt.Println("Time to search slice of 100_000_000 floats using
                    concurrentSearch = ", elapsed)
    fmt.Println("Result of search is ", result)
}
/*
Time to search slice of 100_000_000 floats using concurrentSearch
= 9.666792ms
Result of search is false
Time to search slice of 100_000_000 floats using concurrentSearch
= 5.311917ms
Result of search is true
*/
```

An improvement of over a factor of 5 in worst-case search time is achieved using concurrency. The complexity of the linear search and concurrent search is **O(n)**.

Binary Searches

If the data in the slice to be searched is sorted, a binary search algorithm could be used. This algorithm is **O(log₂n)** since the search space is halved during each iteration.

Listing 2-11 presents the details of this binary search on sorted data.

Listing 2-11. Binary search on sorted data

```
package main

import (
    "fmt"
    "time"
)

const size = 100_000_000

type Ordered interface {
    ~float64 | ~int | ~string
}

func binarySearch[T Ordered](slice []T, target T) bool {
    low := 0
    high := len(slice) - 1

    for low <= high {
        median := (low + high) / 2

        if slice[median] < target {
            low = median + 1
        } else {
            high = median - 1
        }
    }
```

```go
        if low == len(slice) || slice[low] != target {
            return false
        }
        return true
}
func main() {
    data := make([]float64, size)
    for i := 0; i < size; i++ {
        data[i] = float64(i) // is sorted
    }
    start := time.Now()
    result := binarySearch[float64](data, -10.0)
    elapsed := time.Since(start)
    fmt.Println("Time to search slice of 100_000_000 floats using
    binarySearch = ", elapsed)
    fmt.Println("Result of search is ", result)

    start = time.Now()
    result = binarySearch[float64](data, float64(size / 2))
    elapsed = time.Since(start)
    fmt.Println("Time to search slice of 100_000_000 floats using
    binarySearch = ", elapsed)
    fmt.Println("Result of search is ", result)
}
/* Output
Time to search slice of 100_000_000 floats using binarySearch = 1.375µs
Result of search is false
Time to search slice of 100_000_000 floats using binarySearch = 334ns
Result of search is true
*/
```

The search time here is significantly smaller than the search times on random data achieved earlier. That is because the data is sorted.

The fastest sorting algorithm has complexity $O(nlog_2n)$. If one needed to perform many independent searches within the slice, it might pay to sort the data first and then conduct the many searches using a binary search.

It would not be beneficial to sort the slice before performing a single search of the slice because of the overhead of sorting.

2.4 Summary

Big O notation describes the asymptotic efficiency of an algorithm. We examined several classic sorting and searching algorithms in this chapter and characterized them by their big O property. We presented concurrent solutions for each of the sorting algorithms and observed significant improvements in their performance.

In the next chapter, we discuss object-oriented programming in Go without classes.

CHAPTER 3

Abstract Data Types: OOP Without Classes in Go

In the previous chapter, we discussed algorithm complexity and presented examples with and without the use of concurrency.

In this chapter, we show how object-oriented programming can be done without the **class** construct. We review the fundamental concept of abstract data types and illustrate their use with many examples.

3.1 Abstract Data Type Using Classes

In 1980, the **Smalltalk** language, developed at Xerox PARC, was released. This seminal language set the stage for a new paradigm of programming: object-oriented programming. The centerpiece of Smalltalk and newer object-oriented languages that followed is the **class** construct. Some of the major object-oriented languages that followed Smalltalk include Eiffel, C++ (a hybrid language that includes the class construct), Java, Swift, Python, and C#. Each uses the **class** as the central construct for defining abstract data types and describing the behavior of objects that are instances of a class.

The **class** construct implements some older well-established ideas about how software is constructed. Specifically, classes implement **abstract data types**.

An **abstract data type** is characterized by a set of operations that can be performed on the underlying type. Consider the following simple example.

91

© Richard Wiener 2022
R. Wiener, *Generic Data Structures and Algorithms in Go*, https://doi.org/10.1007/978-1-4842-8191-8_3

We define the **Counter** abstract data type as follows:

Attributes

count int – The internal data of each Counter object (instance)

Methods

Increment( ) – Adds one to the current value of the attribute count

Decrement( ) – Subtracts one from the current value of the attribute count only if count > 0

Reset( ) – Sets the value of count to zero

GetCount( ) – Returns the current value of count

If *myCounter* is defined as being of type *Counter* (an instance of some class **Counter**), the following operations would be legal:

```
myCounter.Increment()
myCounter.Decrement()
myCounter.Reset()
countValue = myCounter.GetCount()
```

In the preceding example, *myCounter* is referred to as an object (an instance of class **Counter**), and the method calls connected to each object with the dot operator are the legal operations that could be performed on each object, thus the name **object-oriented programming**.

Object-oriented languages supporting the class construct provide a mechanism for extending the operation set defined in a parent class using inheritance. Languages like C++ and Eiffel allows a subclass to inherit operations from two or more parent classes, while languages such as Java and C# allow inheritance from only one parent class.

Much has been written about inheritance in object-oriented languages. Inheritance has fallen out of favor in recent years because of the complexity it can introduce and the dependencies that may be created in a class hierarchy.

Two recent languages that have abandoned inheritance and in fact have abandoned classes are Go and Rust. But Go and Rust have not abandoned object-oriented programming (OOP) but have changed how this paradigm is used.

Before we delve into the details of OOP in Go, let us examine how we might implement the ***Counter*** abstract data type in Python.

```python
class Counter:

    def __init__(self):
        self.count  = 0

    def increment(self):
        self.count += 1

    def decrement(self):
        self.count -= 1

    def reset():
        self,count = 0

    def get_count(self) -> int:
        return self.count

if __name__ == "__main__":
    my_counter = Counter()
    for index in range(10):
        my_counter.increment()
    my_counter.decrement()
    current_count = my_counter.get_count()
    print(current_count)
''' Output
9
'''
```

The keyword ***self*** is used as a reference to any instance of class ***Counter***. The attribute ***count***, defined in the ***__init__*** method, is stored in each instance (object) of class ***Counter***.

One obvious appeal of the class construct is that all the operations on the underlying attribute(s) are encapsulated in this single construct.

What about Go? In the next section, we look at defining abstract data types in Go without the use of classes.

3.2 Abstract Data Types in Go

Go does not include the class construct. This is a major departure from recent object-oriented languages. Since there is no class construct in Go, there is no inheritance.

ADT Counter

The ADT Counter must restrict the operations that can be performed on an instance of Counter. Specifically, we cannot assign an arbitrary value to a counter. We cannot change the value of a counter by more than one. Without these restrictions, there would be no value in defining this abstract data type. We could use a simple **int** type instead.

Listing 3-1 shows our first implementation of the **Counter** abstract data type (ADT). As we will see shortly, this implementation is faulty. After defining **Counter** as a struct with the field *count*, we define a series of methods that operate on instances, *c*, of **Counter** or pointer to **Counter**.

Listing 3-1. First implementation of Counter ADT

```
package main

import (
    "fmt"
)

type Counter struct {
    count int
}
// Methods
func (c *Counter) Increment() {
    c.count++
}

func (c *Counter) Decrement() {
    c.count--
}
```

```go
func (c *Counter) Reset() {
    c.count = 0
}

func (c Counter) GetCount() int {
    return c.count
}

func main() {
    myCounter := new(Counter)
    // myCounter.count = 100 // Defeats the encapsulatiom of Counter
    fmt.Println(myCounter.GetCount())
    for i := 1; i <= 10; i++ {
        myCounter.Increment()
    }
    myCounter.Decrement()
    // myCounter.count -= 6 // Defeats the encapsulation of Counter
    fmt.Println(myCounter.GetCount())
}
/*
0
9
*/
```

There is a problem with this first implementation. If one were to uncomment the two commented lines of code:

```go
myCounter.count = 100
myCounter.count -= 6
```

the encapsulation that preserves the integrity of the ADT would be broken. The whole point of creating and implementing an ADT is to enforce the abstraction. In this case, that means not allowing the **count** value to be changed by more than one and not allowing the **count** value to be arbitrarily assigned.

We repair the problem and enforce the abstraction, as shown in Listing 3-2. We define a *Counter* interface (using an uppercase C) to formally define the ADT.

Listing 3-2. Second implementation of Counter ADT

```go
// Creating ADT Counter
package main

import (
    "fmt"
)

// This type implicitly implements Counter ADT
type counter struct {
    count int
}

// Interface serves to expose public features of counter
// The attribute count is private
type Counter interface {
    increment()
    decrement()
    reset()
    getCount() int
}

func (c *counter) increment()  {
    c.count += 1
}

func (c *counter) decrement()  {
    if c.count > 0 {
        c.count -= 1
    }
}

func (c *counter) reset() {
    c.count = 0
}

func (c counter) getCount() int {
    return c.count
}
```

```go
func main() {
    myCounter := Counter(&counter{})
    // The only operations that can be performed on myCounter
    // are specified in the Counter interface
    myCounter.increment()
    myCounter.increment()
    myCounter.reset()
    myCounter.increment()
    myCounter.increment()
    myCounter.increment()
    myCounter.increment()
    myCounter.decrement()
    countValue := myCounter.getCount()
    fmt.Println(countValue)
}
// 3
```

The *Counter* interface specifies the signature of the four operations that can be performed on an ADT *Counter*.

The methods *increment()*, *decrement()*, *reset()*, and *getCount()*, each defined on a **counter** type, **implicitly** make **counter** implement the ADT *Counter*.

If we were to comment out the *reset()* method and comment out the *myCounter.reset()* in function **main**, we would get the following compiler error message:

./counter2.go:41:26: cannot convert &counter{} (value of type *counter) to type Counter:

*counter does not implement Counter (missing reset method)

Without the *reset()* method defined on **counter**, the type **counter** no longer can be considered to be of type **Counter**, and the compiler detects this error.

Abstract data types in Go are always **implicitly defined** by defining an interface that specifies the operations associated with the ADT and then implementing methods on the underlying type that have the exact signatures given in the interface specification.

With this accomplished in Listing 3-2, there is no way to violate the ADT encapsulation as was evident in Listing 3-1.

Creating a counter Package

Another way to protect *count* and preserve the encapsulation of the **counter** abstraction is to create a **counter** package and export **Counter** but not **count**.

```
package counter

// Field count is encapsulated as private because it is lowercase
type Counter struct {
    count int // private field
}

func (c *Counter) Increment() {
    c.count += 1
}

func (c *Counter) Decrement() {
    if c.count > 0 {
        c.count -= 1
    }
}

func (c *Counter) Reset() {
    c.count = 0
}

func (c Counter) GetCount() int {
    return c.count
}
```

In package counter, we protect the *count* field of *Counter* from being assigned to outside the package by using a lowercase character as the first character in *count*.

Mechanics of Creating a Package

In order to define a package in a subdirectory of your own choosing, we must follow a set of steps that are outlined in the following. Here, we desire to include the **counter.go** file that defines the counter package in a subdirectory counter in some work directory. A main driver program, **main.go**, is defined in another subdirectory, **maincounter**.

The steps needed to create package counter are the following:

1. Create a subdirectory **counter** in your work directory.

2. Save the **counter.go** file that contains package counter (see the preceding text) in this subdirectory.

3. Create a subdirectory **maincounter** in your work directory.

4. Save the **maincounter.go** file in this subdirectory.

5. Open a terminal window to the **counter** directory.

6. Type the following command: **go mod init example.com/counter**

7. Type the following command: **go mod tidy**

8. Open a terminal window to the maincounter directory.

9. Type the following command: **go mod init example.com/ maincounter**

10. Edit the go.mod file to be

    module example.com/maincounter
    go 1.18

    replace example.com/counter => ../counter

11. Type the following command: **go mod tidy**

Listing 3-3 shows the third implementation of Counter ADT.

Listing 3-3. Third implementation of Counter ADT using package counter

```
// In subdirectory counter
package counter

type Counter struct {
    count int
}

func (c *Counter) Increment()  {
    c.count += 1
}
```

```go
func (c *Counter) Decrement()  {
    if c.count > 0 {
        c.count -= 1
    }
}

func (c *Counter) Reset() {
    c.count = 0
}

func (c Counter) GetCount() int {
    return c.count
}

// In subdirectory maincounter
package main

import (
    "fmt"
    "example.com/counter"
)

func main() {
    myCounter := counter.Counter{}

    myCounter.Increment()
    myCounter.Increment()
    myCounter.Reset()
    myCounter.Increment()
    myCounter.Increment()
    myCounter.Increment()
    myCounter.Increment()
    myCounter.Decrement()
    countValue := myCounter.GetCount()
    fmt.Println(countValue)
}
// 3
```

Identifiers that are exported and therefore available outside of a package must start with an uppercase character. This includes type names such as **Counter** and method names. The field **count** in the **Counter** struct purposely uses a lowercase letter so that its value cannot be accessed outside the package. The only way to change the **count** is through the methods that operate on **Counter**.

Which of the two approaches, given in Listings 3-2 and 3-3, should one use in implementing the ADT **Counter**?

If an ADT is to be used in two or more applications, the solution in Listing 3-3 that defines a package for the ADT is preferred. If an ADT is a one-off, needed only in a specialized application, then the solution using the interface type is easier.

Another Example of Implementing an ADT

We look at another example with more complexity to illustrate how Go implements abstract data types without the use of classes.

Consider the ADT **Employee**.

Attributes

LastName string

FirstName string

Role string

Salary float64

Methods

Get LastName (read-only)

Get FirstName (read-only)

Set/Get Role

Set/Get Salary

String() string – Represents the instance as a string

We have specified **LastName** and **FirstName** as read-only. Once their values have been assigned, they cannot be changed.

Also consider the ADT **PartTimeEmployee** that is an **Employee** with an additional feature.

Attributes

Employee

HourlyWage float64

Methods

Set/Get HourlyWage

String() string – Represents the instance as a string

We deploy the struct and interface definitions shown in the following to establish the ADT:

```go
type employee struct {
    lastName string
    firstName string
    role string
    salary float64
}

type Employee interface {
    SetLastName(lName string)
    SetFirstName(fName string)
    SetRole(r string)
    GetRole() string
    SetSalary(s float64)
    GetSalary() float64
    String() string
}

type partTimeEmployee struct {
    employee
    hourlyWage float64
}
```

```go
type PartTimeEmployee interface {
    Employee
    SetHourlyWage(hourly float64)
    GetHourlyWage() float64
}
```

Using Composition

Each of the struct types, *employee* and *partTimeEmployee*, is accompanied by interface types. These define the operations required on their respective struct types to implicitly make the struct types implement the interfaces given.

We use embedding when we define the first field of *partTimeEmployee* to be *employee*.

In software design, this is called **composition**. The abstraction for a part time employee is composed of an employee and an hourly wage.

The *PartTimeEmployee* interface also uses embedding by including the interface *Employee* first. This requires that all the methods of *Employee* be implemented along with the two new methods, Set/Get hourly wage.

Listing 3-4 fleshes out the ADTs defined previously and presents a short main driver program.

Listing 3-4. Employee and PartTimeEmployee ADTs in action

```go
package main

import (
    "fmt"
)

type employee struct {
    lastName string
    firstName string
    role string
    salary float64
}
```

```go
type Employee interface {
    SetLastName(lName string)
    SetFirstName(fName string)
    SetRole(r string)
    GetRole() string
    SetSalary(s float64)
    GetSalary() float64
    String() string
}

type partTimeEmployee struct {
    employee
    hourlyWage float64
}

type PartTimeEmployee interface {
    Employee
    SetHourlyWage(hourly float64)
    GetHourlyWage() float64
}

// Methods
func (person *employee) SetSalary(yearly float64) {
    person.salary = yearly
}

func (person employee) GetSalary() float64 {
    return person.salary
}

func (person *employee) SetFirstName(firstN string) {
    person.firstName= firstN
}

func (person employee) GetFirstName() string {
    return person.firstName
}
```

```go
func (person *employee) SetLastName(lastN string) {
    person.lastName = lastN
}

func (person *employee) SetRole(r string) {
    person.role = r
}

func (person employee) GetRole() string {
    return person.role
}

func (person employee) String() string {
    result := "Name: " + person.firstName + " " + person.lastName + "\n"
    result += "Role: " + person.role + "\n"
    result += "Annual salary: $" + fmt.Sprintf("%0.2f", person.
    salary) + "\n"
    return result
}

func (person partTimeEmployee) String() string {
    result := "Name: " + person.firstName + " " + person.lastName + "\n"
    result += "Role: " + person.role + "\n"
    result += "HourlyWage: $" + fmt.Sprintf("%0.2f", person.
    hourlyWage) + "\n"
    return result
}

func (person *partTimeEmployee) SetHourlyWage(amt float64) {
    person.hourlyWage = amt
}

func (person partTimeEmployee) GetHourlyWage() float64 {
    return person.hourlyWage
}
```

```go
func main() {
    person := new(employee) // Returns the address of an employee
    person.SetFirstName("Helen")
    person.SetLastName("Rose")
    person.SetRole("Technical Lead")
    person.SetSalary(125_644.0)
    fmt.Println(person.String())

    hourlyWorker := new(partTimeEmployee) // Returns address
    hourlyWorker.SetFirstName("Mark")
    hourlyWorker.SetLastName("Smith")
    hourlyWorker.SetRole("Software Developer")
    hourlyWorker.SetHourlyWage(85.00)
    fmt.Println(hourlyWorker.String())
}
/*
Name: Helen Rose
Role: Technical Lead
Annual salary: $125644.00

Name: Mark Smith
Role: Software Developer
HourlyWage: $85.00
*/
```

Variables *person* and *hourlyWorker* act like objects (instances of a class) in traditional object-oriented programming (OOP) languages. Methods are invoked on these variables as one would do in traditional OOP languages.

In the next section, we discuss polymorphism in Go. This is another fundamental pillar of object-oriented programming.

3.3 Polymorphism

Polymorphism is a basic pillar of object-oriented programming. It allows actions to be taken on objects **at runtime**, where the action is based on the **type of object** that receives the action.

In traditional strongly typed object-oriented languages like C#, Java, and Swift, if the action is declared on a formal type and the actual type is an instance of a descendant class, the runtime system chooses the method belonging to the actual type receiving the message.

This cannot happen in Go since descendant classes (inheritance) do not exist.

Using Interfaces to Achieve Polymorphism

We can achieve polymorphic behavior using interfaces as the next example in Listing 3-5 illustrates.

Listing 3-5. Polymorphism in action

```go
package main

import (
    "fmt"
)

type FixedPriceJob struct {
    description string
    fixedPrice float64
}

type HourlyJob struct {
    description string
    hourlyRate float64
    numberHours int
}

type JobInterface interface {
    Cost() float64
    GetDescription() string
}
```

```go
// Implicitly defines FixedPriceJob as implementing the JobInterface
func (job FixedPriceJob) Cost() float64 {
    return job.fixedPrice
}

func (job FixedPriceJob) GetDescription() string {
    return job.description
}

// Implicitly defines HourlyJob as implementing the JobInterface
func (hourlyJob HourlyJob) Cost() float64 {
    return hourlyJob.hourlyRate * float64(hourlyJob.numberHours)
}

func (hourlyJob HourlyJob) GetDescription() string {
    return hourlyJob.description
}

func TotalJobCost(jobs []JobInterface) float64 {
    result := 0.0
    for _, job := range jobs {
        result += job.Cost()
    }
    return result
}

func main() {
    job1 := FixedPriceJob{"Stucco House", 34760.0}
    job2 := HourlyJob{"Landscaping", 40.0, 50}
    jobs := []JobInterface{job1, job2}
    totalCost := TotalJobCost(jobs)
    fmt.Printf("Total job cost: $%0.2f", totalCost)
}
// Total job cost: $36760.00
```

Any type that defines methods with the signatures given in *JobInterface* implicitly implements this interface. That is what we do in Listing 3-5. We define *Cost()* and *GetDescription()* methods on both *FixedPriceJob* and *HourlyJob*.

In function ***TotalJobCost***, we input a slice of type ***JobInterface***. We iterate over the range of input jobs and accumulate the total cost by invoking the ***Cost()*** method on each job. The runtime system binds the correct ***Cost()*** method based on the type of job receiving this method (whether the job is a **FixedPriceJob** or **HourlyJob** in this example). That is polymorphism in action.

In the next section, we present an object-oriented programming (OOP) application. A simple Blackjack card game is developed.

3.4 OOP Application: Simplified Game of Blackjack

In traditional object-oriented languages such as Smalltalk, Java, C#, and Swift, the design process involves problem decomposition into classes. This is not possible in Go since classes do not exist.

We illustrate how problem decomposition can be achieved in Go. We design and implement a small, simplified Blackjack card game. This game is console based.

In Blackjack, two cards are dealt from the deck to the player and to the house. The goal is to accumulate points but not exceed 21 points. The point value of a card is the number on its face or 10 if the card is a jack, queen, or king or 11 if the card is an ace. If the hand has two or more aces, then 10 is subtracted from the total point count of the hand.

The player goes first and acquires additional cards, if she wishes to, by saying "hit me." When the player's score gets close to 21, the player stops. If the player's score exceeds 21 after being "hit," the game ends with the house as the winner. If not, it is the house's turn. Here, we simplify things by assuming that the house will request "hit me" if its total score is less than 17. After the house is finished with its play, the winner is the one with the highest score if that score is less than or equal to 21. Ties are possible.

In traditional object-oriented languages, we would define classes Card, Hand, and Deck and also define methods for taking actions on these entities.

In Go, we model the system using structs and methods.

Consider the following:

```
var ranks = []string {"2", "3", "4", "5", "6", "7", "8", "9", "10", "J",
            "Q", "K", "A"}
var suits = []rune {'\u2660', '\u2661', '\u2662', '\u2663'}
```

```go
type Card struct {
    Rank string
    Suit string
}

type Hand struct {
    Cards []Card
}

type Deck struct {
    Cards []Card
}
```

Variable *ranks* is a slice containing the available cards, each a string. Variable *suits* contains a slice of four rune values representing the symbols for club, diamond, heart, and spade.

Type *Card* is a struct with the fields *Rank* and *Suit*.

The method *value* operates on a *hand* as follows:

```go
func (hand Hand) value() int {
    result := 0
    numberAces := 0
    for index := 0; index < len(hand.Cards); index++ {
        if hand.Cards[index].Rank != "A" && hand.Cards[index].Rank
        != "K" &&
                    hand.Cards[index].Rank != "Q" && hand.Cards[index].
                    Rank != "J" {
            intVal, _ := strconv.Atoi(hand.Cards[index].Rank)
            result += intVal
        } else if hand.Cards[index].Rank == "J" || hand.Cards[index].
        Rank == "Q" ||
                    hand.Cards[index].Rank == "K" {
            result += 10
        } else if hand.Cards[index].Rank == "A"{
            result += 11
            numberAces += 1
        }
    }
```

```go
    if result > 21 && numberAces > 1 {
        result -= 10 * numberAces
    }
    return result
}
```

The other supporting methods are presented in the following:

```go
func (hand *Hand) addCard(card Card) {
    hand.Cards = append(hand.Cards, card)
}

func (hand Hand) Display() {
    fmt.Println("\n")
    for _, card := range hand.Cards {
        fmt.Print(card.Rank + card.Suit + " ")
    }
}

func (deck *Deck) dealCard() Card {
    result := deck.Cards[0]
    deck.Cards = deck.Cards[1:]
    return result
}

func (deck *Deck) shuffle() {
    rand.Seed(time.Now().UnixNano())
    rand.Shuffle(len(deck.Cards), func(i, j int) { deck.Cards[i],
                    deck.Cards[j] = deck.Cards[j], deck.Cards[i]
    })
}

func (deck *Deck) initializeDeck() Deck{
    for _, suit := range suits {
        for _, rank := range ranks {
            deck.Cards = append(deck.Cards, Card{rank, string(suit)})
        }
    }
```

```go
    deck.shuffle()
    return *deck
}

func (deck Deck) display() {
    for _, card := range deck.Cards {
        fmt.Print(card.Rank + card.Suit + " ")
    }
}
```

Method **shuffle** utilizes the Shuffle function from package "math/rand".

Listing 3-6 presents the complete Go application for Blackjack.

Listing 3-6. Blackjack

```go
package main

import (
    "strconv"
    "fmt"
    "math/rand"
    "time"
    "bufio"
    "os"
)

var ranks = []string {"2", "3", "4", "5", "6", "7", "8", "9", "10", "J",
            "Q", "K", "A"}
var suits = []rune {'\u2660', '\u2661', '\u2662', '\u2663'}

type Card struct {
    Rank string
    Suit string
}

type Hand struct {
    Cards []Card
}
```

```go
type Deck struct {
    Cards []Card
}

func (hand Hand) value() int {
    result := 0
    numberAces := 0
    for index := 0; index < len(hand.Cards); index++ {
        if hand.Cards[index].Rank != "A" && hand.Cards[index].
        Rank != "K" &&
                    hand.Cards[index].Rank != "Q" && hand.Cards[index].
                    Rank != "J" {
            intVal, _ := strconv.Atoi(hand.Cards[index].Rank)
            result += intVal
        } else if hand.Cards[index].Rank == "J" || hand.
        Cards[index].Rank == "Q" ||
                    hand.Cards[index].Rank == "K" {
            result += 10
        } else if hand.Cards[index].Rank == "A"{
            result += 11
            numberAces += 1
        }
    }
    if result > 21 && numberAces > 1 {
        result -= 10 * numberAces
    }
    return result
}

func (hand *Hand) addCard(card Card) {
    hand.Cards = append(hand.Cards, card)
}
```

```go
func (hand Hand) Display() {
    fmt.Println("\n")
    for _, card := range hand.Cards {
        fmt.Print(card.Rank + card.Suit + " ")
    }
}

func (deck *Deck) dealCard() Card {
    result := deck.Cards[0]
    deck.Cards = deck.Cards[1:]
    return result
}

func (deck *Deck) shuffle() {
    rand.Seed(time.Now().UnixNano())
    rand.Shuffle(len(deck.Cards), func(i, j int) { deck.Cards[i],
                        deck.Cards[j] = deck.Cards[j], deck.Cards[i] })
}

func (deck *Deck) initializeDeck() Deck{
    for _, suit := range suits {
        for _, rank := range ranks {
            deck.Cards = append(deck.Cards, Card{rank, string(suit)})
        }
    }
    deck.shuffle()
    return *deck
}

func (deck Deck) display() {
    for _, card := range deck.Cards {
        fmt.Print(card.Rank + card.Suit + " ")
    }
}
```

```go
func main() {
    gameOver := false
    myDeck := Deck{}
    myDeck.initializeDeck()
    houseHand := Hand{}
    playerHand := Hand{}
    for i := 1; i <= 2; i++ {
        card := myDeck.dealCard()
        houseHand.addCard(card)
        card = myDeck.dealCard()
        playerHand.addCard(card)
    }
    playerHand.Display()
    fmt.Println("    Do you want to be hit (y/n)?")
    reader := bufio.NewReader(os.Stdin)
    res, _ , _:= reader.ReadRune()
    for ; ; {
        if res != 'y' {
            break
        }
        card := myDeck.dealCard()
        playerHand.addCard(card)
        playerHand.Display()
        if  playerHand.value() > 21 {
            fmt.Println("PLAYER'S SCORE EXCEEDS 21.  GAME OVER. HOUSE WINS!")
            gameOver = true
            break
        }
        fmt.Println("    Do you want to be hit (y/n)?")
        reader = bufio.NewReader(os.Stdin)
        res, _ , _ = reader.ReadRune()
    }
    if !gameOver {
        for ; ; {
            if houseHand.value() > 21 {
```

```go
            fmt.Println("HOUSE SCORE EXCEEDS 21. GAME OVER. PLAYER WINS!")
            gameOver = true
            break
        }
        if houseHand.value() < 17 {
            card := myDeck.dealCard()
            houseHand.addCard(card)
        } else {
            break
        }
    }
}
if !gameOver {
    if playerHand.value() > houseHand.value() {
        fmt.Println("PLAYER SCORE EXCEEDS HOUSE SCORE.  GAME OVER.
                  PLAYER WINS!")
    } else if playerHand.value() == houseHand.value() {
        fmt.Println("PLAYER SCORE EQUALS HOUSE SCORE.  GAME OVER.
                  TIE GAME!")
    } else {
        fmt.Println("HOUSE SCORE EXCEEDS PLAYER SCORE.  GAME OVER.
        HOUSE                     WINS!")
    }
}
}
```

The output of one of many runs is

4♡ 3♣ Do you want to be hit (y/n)?

y

4♡ 3♣ 6♠ Do you want to be hit (y/n)?

y

4♡ 3♣ 6♠ 5♡ Do you want to be hit (y/n)?

n

HOUSE SCORE EXCEEDS PLAYER SCORE. GAME OVER. HOUSE WINS!

In the final section of this chapter, we present another OOP application. This application utilizes the standard **map** data structure defined in Go.

3.5 Another OOP Application: Permutation Group of Words

A permutation group of words contains a collection of words that are formed from the same letters and are all found in the same dictionary.

For example, a permutation group for "persist" contains a collection of words that are formed from the same letters and are all found in the same dictionary.

The permutation group for "persist" is ['esprits', 'persist', 'priests', 'spriest', 'sprites', 'stirpes', 'stripes'].

One's first thought might be to enumerate all permutations of the group of letters and see what subset is in the dictionary.

Using the Standard map Data Structure

We will take a different approach. As we scan an entire file of words, we construct a **map** with **key-value** pairs as follows:

key: Alphabetized word (all the letters of the given word rearranged from smallest letter to largest letter). For example, alphabetized("camp") = "acmp", alphabetized("balloon") = "abllnoo"

value: A collection of dictionary words that can be reduced to the same alphabetized word

As we process each word in a **words.txt** file, we compute the key by alphabetizing the word and then check to see whether the key is already present in our map. If it is, we add the word we are processing to the value collection associated with this key. If not, we create a new collection and add the <alphabetized(word), new collection> key-value pair to our map.

When the map is done, we find the permutations of a specified word by computing its key and then getting the collection associated with this key from our map.

We start by defining a global variable ***dictionary***.

```
var dictionary map[string][]string
```

117

Next, we define a function, ***alphabetize***.

```go
func alphabetize(word string) string {
    s := strings.Split(word, "")
    sort.Strings(s)
    return strings.Join(s, "")
}
```

The first line creates an array of characters. The next line sorts this array in place. The third line joins the sorted array forming the resulting string.

The function, ***buildDictionary***, creates a map with each key representing a sorted alphabetized word and each value being a slice of words that alphabetize to the key.

This function is shown next.

```go
func buildDictionary() {
    dictionary = make(map[string][]string)

    file, err := os.Open("words.txt")

    if err != nil {
        log.Fatalf("failed opening file: %s", err)
    }

    scanner := bufio.NewScanner(file)
    scanner.Split(bufio.ScanLines)
    var txtwords []string

    for scanner.Scan() {
        txtwords = append(txtwords, scanner.Text())
    }

    file.Close()

    for _, word := range txtwords {
        alphabetized := alphabetize(word)
        var lst []string
        if len(dictionary) > 0 && len(dictionary[alphabetized]) > 0 {
            lst = dictionary[alphabetized]
```

```go
    } else {
        lst = []string{}
    }
    lst = append(lst, word)
    dictionary[alphabetized] = lst
    }
}
```

The file handling portion of **buildDictionary** is the most complex.

```go
file, err := os.Open("words.txt")

if err != nil {
    log.Fatalf("failed opening file: %s", err)
}

scanner := bufio.NewScanner(file)
scanner.Split(bufio.ScanLines)
var txtwords []string
for scanner.Scan() {
    txtwords = append(txtwords, scanner.Text())
}

file.Close()
```

Using the imported package, *os*, a text file of words is opened. A *scanner* is defined by using *NewScanner* on the *bufio* package that is imported. Using *scanner.Scan()*, the slice of words contained in the **words.txt** file is generated.

Each word in this slice is alphabetized and either added to the existing map for that key or a new key is created and the first value in the slice of words associated with the key is inserted.

Listing 3-7 presents the full source code for this application.

Listing 3-7. Permutation Group of Words

package main

```go
import (
    "fmt"
    "sort"
```

```go
    "strings"
    "bufio"
    "log"
    "os"
)

func init() {
    buildDictionary()
}

var dictionary map[string][]string

func alphabetize(word string) string {
    s := strings.Split(word, "")
    sort.Strings(s)
    return strings.Join(s, "")
}

func buildDictionary() {
    dictionary = make(map[string][]string)

    file, err := os.Open("words.txt")

    if err != nil {
        log.Fatalf("failed opening file: %s", err)
    }

    scanner := bufio.NewScanner(file)
    scanner.Split(bufio.ScanLines)
    var txtwords []string

    for scanner.Scan() {
        txtwords = append(txtwords, scanner.Text())
    }

    file.Close()

    for _, word := range txtwords {
        alphabetized := alphabetize(word)
        var lst []string
        if len(dictionary) > 0 && len(dictionary[alphabetized]) > 0 {
```

```go
            lst = dictionary[alphabetized]
        } else {
            lst = []string{}
        }
        lst = append(lst, word)
        dictionary[alphabetized] = lst
    }
}

func output(word string) {
    wd := alphabetize(word)
    fmt.Printf("Permutation group of %s is %s", word, dictionary[wd])
}

func main() {
    output("parties")
}
// Permutation group of parties is [parties pastier piaster piastre pirates
// raspite spirate tapiser traipse]
```

3.6 Summary

We focused on the implementation of abstract data types in this chapter. Two
approaches were shown to accomplish this. The first uses an interface to define the
required operations given by the ADT. The second uses a package to expose the
public features required by the ADT while hiding internal features. We introduced the
important concept of polymorphism. This allows the runtime system to determine
which particular method to bind to an object receiving the method assuming that the
object is of a type implementing the interface. We presented several examples of object-
oriented programming.

   In the next chapter, we present a larger example of object-oriented programming by
showing an implementation of the Game of Life. We utilize a third-party graphical user
interface (GUI) package.

ADT in Action: Game of Life

In the previous chapter, we showed how abstract data types can be implemented and how object-oriented programming can be performed in Go. In this chapter, we continue to explore object-oriented programming in Go. We implement the classic Game of Life. We introduce and utilize a third-party GUI package as part of our implementation.

In the next section, we specify the Game of Life.

4.1 Game

To illustrate the central role that ADTs can play in software design, we explore the Game of Life, invented by John Conway and published in 1970 by *Scientific American*. This game is a cellular automaton and interesting to design, implement, and observe.

In addition to showcasing the central role of an ADT in the design of this game, we introduce the **fyne** graphical user interface (GUI) framework in Go.

We start with an empty grid with R rows and C columns. Clusters of live cells are created at random locations. Then the internal rules of grid evolution take over, and the user can observe each successive grid evolution at one-second intervals.

Rules of Grid Cell Evolution

The rules of grid cell evolution to produce the next generation of grid cells are the following:

1. Any live cell that has zero or one neighbor dies (disappears from the grid in the next generation).

2. Any live cell with four or more neighboring live cells dies (disappears from the grid in the next generation).

© Richard Wiener 2022
R. Wiener, *Generic Data Structures and Algorithms in Go*, https://doi.org/10.1007/978-1-4842-8191-8_4

3. Any live cell with two or three neighboring live cells survives to the next generation.

4. Any empty cell with exactly three live neighbors becomes a live cell in the next generation.

Let us consider the evolution of the game starting with Figure 4-1.

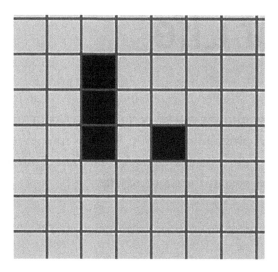

Figure 4-1. *Initial configuration*

The next iteration evolves into Figure 4-2.

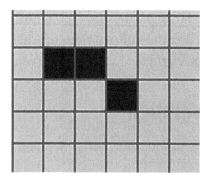

Figure 4-2. *First iteration*

Here, two cells are brought to life (rightmost cell and leftmost cell), and one cell survives (second rightmost cell). The other cells die.

Then the next iteration evolves into Figure 4-3.

Figure 4-3. *Second iteration*

And finally on the last iteration, the configuration evolves into Figure 4-4.

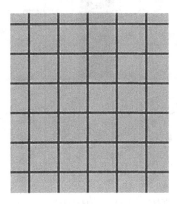

Figure 4-4. *Final iteration*

Very interesting patterns emerge as the Game of Life evolves. Sometimes, oscillations occur forever. Figures 4-5 and 4-6 provide an example.

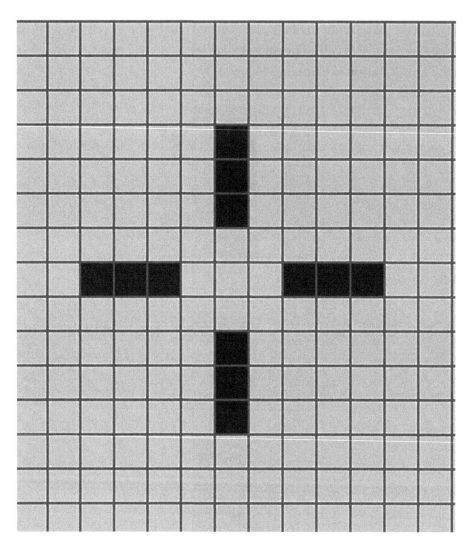

Figure 4-5. *Initial configuration for oscillation*

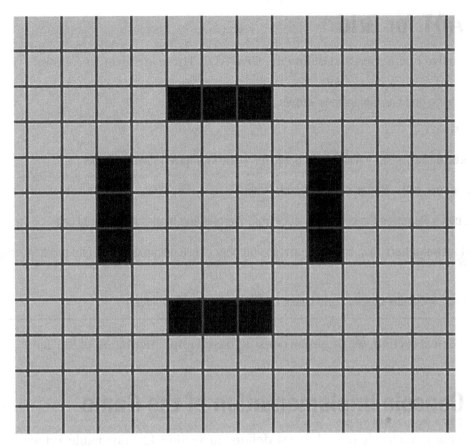

Figure 4-6. *Final configuration for oscillation*

The pattern jumps back and forth from the first pattern to the second pattern as the game evolves.

In the next section, we define an abstract data type (ADT) for grid.

4.2 ADT for Grid

There are five operations that define the ***Grid*** ADT. These are given as follows:

The underlying grid has dimensions **<rows, cols>**.

<u>Operations</u>

initializeGrid(rows, cols) – Allocates storage for a grid with given rows and cols

bringAlive(row, col) – Brings cell <row, col> to life

kill(row, col) – Removes the cell <row, col> from the grid and makes it an empty cell

numberLiveNeighbors(row, col) – Returns the number of live neighbors from grid position <row, col>

evolveGrid() – Obtains the next grid based on the four rules of evolution

---

In the next section, we present a console-based implementation of the game.

4.3 Console Implementation of the Game

In this section, we implement the ADT defined in Section 4.2 and enable a stepwise console output.

The ADT defined is implemented using the following methods with ***g*** of type ***Grid***, the receiver:

```
type Grid [][]bool

func (g *Grid) initializeGrid(r, c int)

func (g Grid) bringAlive(row, col int)

func (g Grid) kill(row, col int)

func (g Grid) numberLiveNeighbors(row, col int) int

func (g Grid) evolveGrid()
```

The method *initializeGrid* is implemented as follows:

```
func (g *Grid) initializeGrid(r, c int) {
    rows = r
    cols = c
    *g = make([][]bool, rows)
    for row := 0; row < rows; row++ {
        (*g)[row] = make([]bool, cols)
    }
}
```

The global variables *rows* and *cols* are assigned to the input parameters *r* and *c*. Storage is allocated to hold *rows* of data. For each *row*, storage is allocated to hold *cols* of data.

Because the receiver is a pointer to **Grid**, the receiver, **grid**, is initialized in place.

The function *numberLiveNeighbors* is implemented in the following. Although the details are somewhat tedious, they are straightforward.

```
func (g Grid) numberLiveNeighbors(row, col int) int {
    result := 0
    if row > 0 && g[row - 1][col] == true {
        result++
    }
    if row > 0 && col < cols - 1 && g[row - 1][col + 1]
            == true {
        result += 1
    }
    if col < cols - 1 && g[row][col + 1] == true {
        result += 1
    }
    if row < rows - 1 && col < cols -1
                && g[row + 1][col + 1] == true {
        result += 1
    }
    if row < rows - 1 && g[row + 1][col] == true {
        result += 1
    }
```

```
    if row < rows - 1 && col > 0 &&
                g[row + 1][col - 1] == true {
        result += 1
    }
    if col > 0 && g[row][col - 1] == true {
        result += 1
    }
    if row >  0 && col > 0 &&
            g[row - 1][col - 1] == true {
        result += 1
    }
    return result
}
```

The method *evolveGrid* implements the business logic – the four rules that specify how the game evolves. This method is implemented as follows:

```
func (g Grid) evolveGrid() {
    Copy(newGrid, g)
    for row := 0; row < rows; row++ {
        for col := 0; col < cols; col++ {
            liveN := g.numberLiveNeighbors(row, col)
            // Rules 1 and 2
            if g[row][col] == true && (liveN < 2 ||
                        liveN >= 4) {
                newGrid[row][col] = false
            }
            // Rule 4
            if g[row][col] == false && liveN == 3 {
                newGrid[row][col] = true
            }
        }
    }
    Copy(g, newGrid)
}
```

A locally created ***newGrid*** is used and initialized to the receiver, **g**. The number of live neighbors is computed, and in the next two **if** clauses, the rules for a live cell being killed or an empty cell coming alive are exercised.

At the end, the locally created ***newGrid*** is copied back to the receiver, **g**.

Listing 4-1 puts the pieces together along with a main driver and shows the output for a specified input.

Listing 4-1. Console implementation of the Game of Life

```go
package main

import (
    "fmt"
    "time"
)

var (
    rows int
    cols int
)

type Grid [][]bool

var grid Grid
var newGrid Grid

func (g *Grid) initializeGrid(r, c int) {
    rows = r
    cols = c
    *g = make([][]bool, rows)
    for row := 0; row < rows; row++ {
        (*g)[row] = make([]bool, cols)
    }
}

func Copy(target [][]bool, source [][]bool) {
    for row := 0; row < rows; row++ {
        for col := 0; col < cols; col++ {
```

```go
            target[row][col] = source[row][col]
        }
    }
}

func (g Grid) bringAlive(row, col int) {
    g[row][col] = true
}

func (g Grid) kill(row, col int) {
    g[row][col] = false
}

func (g Grid) numberLiveNeighbors(row, col int) int {
    result := 0
    if row > 0 && g[row - 1][col] == true {
        result++
    }
    if row > 0 && col < cols - 1 && g[row - 1][col + 1] == true {
        result += 1
    }
    if col < cols - 1 && g[row][col + 1] == true {
        result += 1
    }
    if row < rows - 1 && col < cols -1 && g[row + 1][col + 1] == true {
        result += 1
    }
    if row < rows - 1 && g[row + 1][col] == true {
        result += 1
    }
    if row < rows - 1 && col > 0 && g[row + 1][col - 1] == true {
        result += 1
    }
    if col > 0 && g[row][col - 1] == true {
        result += 1
    }
```

```
    if row > 0 && col > 0 && g[row - 1][col - 1] == true {
        result += 1
    }
    return result
}

func (g Grid) evolveGrid() {
    Copy(newGrid, g)
    for row := 0; row < rows; row++ {
        for col := 0; col < cols; col++ {
            liveN := g.numberLiveNeighbors(row, col)
            // Rules 1 and 2
            if g[row][col] == true && (liveN < 2 || liveN >= 4) {
                newGrid[row][col] = false
            }
            // Rule 4
            if g[row][col] == false && liveN == 3 {
                newGrid[row][col] = true
            }
        }
    }
    Copy(g, newGrid)
}

func consoleOutput() {
    for row := 0; row < rows; row++ {
        for col := 0; col < cols; col++ {
            if grid[row][col] == true {
                fmt.Print("$ ")
            } else {
                fmt.Print("# ")
            }
        }
        fmt.Print("\n")
    }
    fmt.Println("-----")
}
```

```
func main() {
    grid.initializeGrid(3, 3)
    newGrid.initializeGrid(3, 3)

    grid.bringAlive(0, 0)
    grid.bringAlive(0, 2)
    grid.bringAlive(1, 0)
    grid.bringAlive(1, 1)
    grid.bringAlive(2, 2)
    consoleOutput()

    for iteration := 1; iteration < 5; iteration++ {
        time.Sleep(1 * time.Second)
        grid.evolveGrid()
        consoleOutput()
    }
}
/* Output
$ # $
$ $ #
# # $
-----
$ # #
$ # $
# $ #
-----
# $ #
$ # #
# $ #
-----
# # #
$ $ #
# # #
-----
# # #
```

```
# # #
# # #
- - - - -
*/
```

In the text-based console output, dollar signs, $, are used to represent live cells, and pound symbols, #, are used to represent empty cells.

A new grid is displayed every second.

Let us carefully examine the evolution from the initial state to the next state.

```
$ # $
$ $ #
# # $
- - - - -
$ # #
$ # $
# $ #
```

The live cell at <0, 0> survives since it has two live neighbors.

The empty cell at <0, 1> remains empty since it has four live neighbors.

The live cell at <0, 2> does not survive since it has only one live neighbor.

The live cell at <1, 0> survives since it has two live neighbors.

The live cell at <1, 1> does not survive since it has four live neighbors.

The empty cell at <1, 2> comes alive since it has three live neighbors.

The empty cell at <2, 0> remains empty since it has two live neighbors.

The empty cell at <2, 1> comes alive since it has three live neighbors.

And finally, the live cell at <2,2> does not survive since it has one live neighbor.

It is left to the reader to verify that the remaining three grids correctly follow the rules of evolution.

In the next section, we implement a GUI version of the game.

4.4 GUI Implementation of the Game of Life

Applications that require graphical user interfaces (GUIs) in Go are dependent on third-party libraries since there are no built-in GUI libraries. One such third-party library that we shall use here and in later chapters is the **Fyne** library.

A reference on the Fyne library is the book by Andrew Williams: *Building Cross-Platform GUI Applications with Fyne and the Go Programming Language*, Packt Publishing, 2021.

Listing 4-2 presents a GUI solution to the Game of Life.

Listing 4-2. GUI version of the Game of Life

```go
package main

import (
    "math/rand"
    "time"
    "image/color"
    "fyne.io/fyne/v2"
    "fyne.io/fyne/v2/app"
    "fyne.io/fyne/v2/canvas"
    "fyne.io/fyne/v2/container"
)

var (
    rows int
    cols int
    rect *canvas.Rectangle
    // Holds rectangle objects
    segments = []fyne.CanvasObject
)
// Snip from Listing 4.1

func output() *fyne.Container {
    for row := 0; row < rows; row++ {
        for col := 0; col < cols; col++ {
            if grid[col][row] == false {
                rect =
                    canvas.NewRectangle(&color.RGBA{B:
                        200, R: 200, G:200, A: 255})
            } else {
                rect =
```

```
                 canvas.NewRectangle(&color.RGBA{B:
                         0, R: 255, G: 0, A: 255})
            }
            rect.Resize(fyne.NewSize(10, 10))
            rect.Move(fyne.NewPos(float32(row * 11),
                    float32(col * 11)))
            segments = append(segments, rect)
        }
    }
    return container.NewWithoutLayout(segments...)
}

func main() {
    grid.initializeGrid(25, 25)
    newGrid.initializeGrid(25, 25)

    for numberCritters := 0; numberCritters < 4;
                numberCritters++ {
        r := 5 + rand.Intn(10)
        c := 5 + rand.Intn(10)
        grid.bringAlive(r, c)
        grid.bringAlive(r + 1, c)
        grid.bringAlive(r + 1, c + 1)
        grid.bringAlive(r - 1, c)
        grid.bringAlive(r - 2, c - 1)
    }

    a := app.New()
    w := a.NewWindow("GAME OF LIFE - Hit Any Key To
                        Quit")
    w.Resize(fyne.NewSize(300, 300))
    w.SetFixedSize(true)

    go func() {
        for ; ; {
            container := output()
            w.SetContent(container)
```

```
            time.Sleep(1 * time.Second)
            grid.evolveGrid()
        }
    }()

    w.Canvas().SetOnTypedKey(func(k *fyne.KeyEvent) {
        // Shuts down simulation
        w.Close()
    })

    w.ShowAndRun()
}
```

Function *output* returns a **fyne** container. This container contains a grid of colored 10 x 10 rectangles based on whether **grid[row][col]** is true or false.

In *main*, four clusters of live cells are created at random locations. A new **fyne** window is created and sized at 300 × 300 pixels.

In a goroutine, the content of the container is displayed on the **fyne** window every second. The content is changed by the method *evolveGrid()*. The output keeps evolving until the user presses any key. This action closes the window and terminates the program.

Creating go.mod file

For your program to access the myriad of functions imported from the **fyne** library, you need to create a **go.mod** file as follows:

(1) **go mod init guigameoflife.go**

(2) **go mod tidy**

These two commands, executed from a terminal window containing the program, produce the needed **go.mod** and **sum.mod** files needed for program execution.

Program Output

Two screenshots taken during the evolution of the game are shown in the following. The second screen shot shows a steady-state unchanging pattern. This often happens.

Beautiful patterns evolve as the game progresses (Figures 4-7 and 4-8).

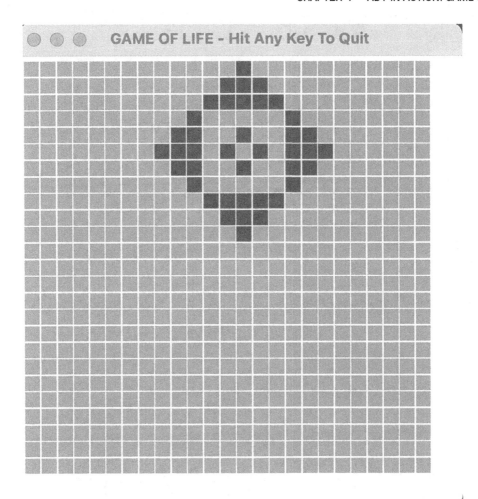

Figure 4-7. *Pattern during game evolution*

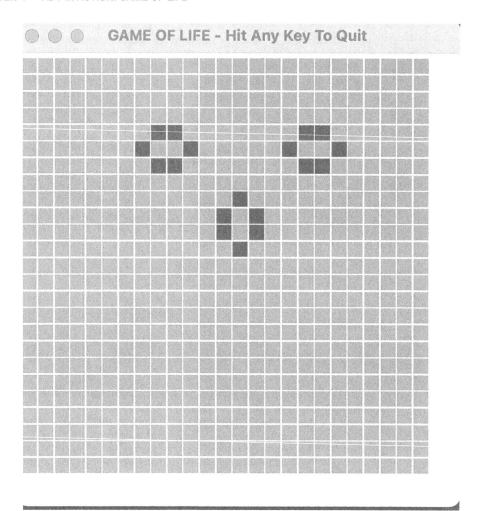

Figure 4-8. *Steady-state pattern*

4.5 Summary

In this chapter, we presented a console-based and GUI-based implementation of the Game of Life. We defined an ADT based on the rules of evolution in the game specification. We used a third-party GUI package to depict the grid and its cells.

In the next chapter, we start the data structure portion of this book. We focus on **Stack** and present some generic stack implementations along with some applications that use a stack.

CHAPTER 5

Stacks

The previous chapter presented an application of abstract data types, the Game of Life.

In this chapter, we switch gears and begin our exploration of generic data structures. The first and perhaps simplest data structure we look at is the **Stack**. It has many practical uses in application development.

A stack organizes data in a last-in, first-out (LIFO). Only the last item inserted into a stack is accessible.

Because of LIFO, the most obvious application is to reverse a sequence of insertions. For example, if the items in a list are inserted onto a stack, a new list that is the reverse of the original list may be obtained by successively popping the elements of the stack.

In the next section, we formalize the Stack abstract data type.

5.1 Stack ADT

There are four operations that characterize a Stack ADT.

Push(item) – Adds item to the stack

Pop( ) item – Removes and returns the last item pushed onto the stack

Top( ) item – Accesses the last item pushed onto the stack without altering the stack

IsEmpty bool – Returns true if the stack has no items, otherwise returns false

---

The first implementation of a stack that we consider is presented in the next section where we consider a slice implementation.

© Richard Wiener 2022
R. Wiener, *Generic Data Structures and Algorithms in Go*, https://doi.org/10.1007/978-1-4842-8191-8_5

5.2 Slice Implementation of Generic Stack

The first implementation of generic stack, presented in Listing 5-1, uses a slice to hold the data in the stack.

Listing 5-1. Slice implementation of generic stack

```go
package main

import (
    "fmt"
)

type Ordered interface {
    ~float64 | ~int | ~string
}

type Stack[T Ordered] struct {
    items []T
}

func getZero[T Ordered]() T {
    var result T
    return result
}

// Methods
func (stack *Stack[T]) Push(item T) {
    // item is added to the right-most position in the
    // slice
    if item != getZero[T]() { // We exclude item if it
                              // is getZero[T]()
        stack.items = append(stack.items, item)
    }
}

func (stack *Stack[T]) Pop() T {
    length := len(stack.items)
    if length > 0 {
```

```
        returnValue := stack.items[length - 1]
        stack.items = stack.items[:(length - 1)]
        return returnValue
    } else {
        return getZero[T]()
    }
}

func (stack Stack[T]) Top() T {
    length := len(stack.items)
    if length > 0 {
        return stack.items[length - 1]
    } else {
        return getZero[T]()
    }
}

func (stack Stack[T]) IsEmpty() bool {
    return len(stack.items) == 0
}

func main() {
    // Create a stack of names
    nameStack := Stack[string]{}
    nameStack.Push("Zachary")
    nameStack.Push("Adolf")
    topOfStack := nameStack.Top()
    if topOfStack != getZero[string]() {
        fmt.Printf("\nTop of stack is %s", topOfStack)
    }
    poppedFromStack := nameStack.Pop()
    if poppedFromStack != getZero[string]() {
        fmt.Printf("\nValue popped from stack is %s",
                        poppedFromStack)
    }
    poppedFromStack = nameStack.Pop()
        if poppedFromStack != getZero[string]() {
```

```go
        fmt.Printf("\nValue popped from stack is %s",
                        poppedFromStack)
    }
    poppedFromStack = nameStack.Pop()
    if poppedFromStack != getZero[string]() {
        fmt.Printf("\nValue popped from stack is %s",
                        poppedFromStack)
    }
    poppedFromStack = nameStack.Pop()
    if poppedFromStack != getZero[string]() {
        fmt.Printf("\nValue popped from stack is %s",
                        poppedFromStack)
    }

    // Create a stack of integers
    intStack := Stack[int]{}
    intStack.Push(5)
    intStack.Push(10)
    intStack.Push(0) // Problem since 0 is the zero
                     // value for int
    top := intStack.Top()
    if top != getZero[int]() {
        fmt.Printf("\nValue on top of intStack is %d", top)
    }
    popFromStack := intStack.Pop()
    if popFromStack != getZero[int]() {
        fmt.Printf("\nValue popped from intStack is
                    %d", popFromStack)
    }
    popFromStack = intStack.Pop()
    if popFromStack != getZero[int]() {
        fmt.Printf("\nValue popped from intStack is
                %d", popFromStack)
    }
    popFromStack = intStack.Pop()
```

```
    if popFromStack != getZero[int]() {
        fmt.Printf("\nValue popped from intStack is
                %d", popFromStack)
    }
}
}
/* Output
Top of stack is Adolf
Value popped from stack is Adolf
Value popped from stack is Zachary
Value on top of intStack is 10
Value popped from intStack is 10
Value popped from intStack is 5
*/
```

The Get Zero Function

The function ***getZero[T]()*** returns a "zero value" associated with the generic parameter, **T**. This special value is returned from the functions **Pop()** and **Top()** if the slice, ***items***, contained within the stack is empty.

Since we are using the "zero value" as a sentinel, indicating an empty stack, we cannot allow this "zero value" to be pushed onto the stack.

Why T Is Declared As Ordered

If you are wondering why we require **T** to be **Ordered**, rather than **any**, consider the statement ***if item != getZero[T]()*** in method ***Push***. The generic type, **T**, must be **Ordered** for this statement to be valid. That is, we need to be assured that two variables of type **T** can be compared. This is an unfortunate requirement fostered by this implementation since there is nothing intrinsic about the stack abstraction that requires the data being held to be ordered.

When we create a stack of integers, the third value we push, value 0, is blocked from insertion onto the stack because it happens to be the "zero value" of type **int**.

So this first implementation of generic stack using a slice to hold the data is seriously flawed.

We examine a second implementation in Listing 5-2.

Listing 5-2. Another slice implementation of generic stack

```go
package main

import (
    "fmt"
)

type Stack[T any] struct {
    items []T
}

// Methods
func (stack *Stack[T]) Push(item T) {
    // item is added to the right-most position in the
    // slice
    stack.items = append(stack.items, item)
}

func (stack *Stack[T]) Pop() T {
    length := len(stack.items)
    returnValue := stack.items[length - 1]
    stack.items = stack.items[:(length - 1)]
    return returnValue
}

func (stack Stack[T]) Top() T {
    length := len(stack.items)
    return stack.items[length - 1]
}

func (stack Stack[T]) IsEmpty() bool {
    return len(stack.items) == 0
}

func main() {
    // Create a stack of names
    nameStack := Stack[string]{}
    nameStack.Push("Zachary")
```

```
nameStack.Push("Adolf")

if !nameStack.IsEmpty() {
    topOfStack := nameStack.Top()
    fmt.Printf("\nTop of stack is %s", topOfStack)
}

if !nameStack.IsEmpty() {
    poppedFromStack := nameStack.Pop()
    fmt.Printf("\nValue popped from stack is %s",
                poppedFromStack)
}

if !nameStack.IsEmpty() {
    poppedFromStack := nameStack.Pop()
    fmt.Printf("\nValue popped from stack is %s",
                poppedFromStack)
}

if !nameStack.IsEmpty() {
    poppedFromStack := nameStack.Pop()
    fmt.Printf("\nValue popped from stack is %s",
                poppedFromStack)
}

if !nameStack.IsEmpty() {
    poppedFromStack := nameStack.Pop()
    fmt.Printf("\nValue popped from stack is %s",
                poppedFromStack)
}

// Create a stack of integers
intStack := Stack[int]{}
intStack.Push(5)
intStack.Push(10)
intStack.Push(0)
```

```
    if !intStack.IsEmpty() {
        top := intStack.Top()
        fmt.Printf("\nValue on top of intStack is %d", top)
    }

    if !intStack.IsEmpty() {
        popFromStack := intStack.Pop()
        fmt.Printf("\nValue popped from intStack is
                        %d", popFromStack)
    }

    if !intStack.IsEmpty() {
        popFromStack := intStack.Pop()
        fmt.Printf("\nValue popped from intStack is
                        %d", popFromStack)
    }

    if !intStack.IsEmpty() {
        popFromStack := intStack.Pop()
        fmt.Printf("\nValue popped from intStack is
                        %d", popFromStack)
    }
}
/* Output
Top of stack is Adolf
Value popped from stack is Adolf
Value popped from stack is Zachary
Value on top of intStack is 0
Value popped from intStack is 0
Value popped from intStack is 10
Value popped from intStack is 5
*/
```

In this second implementation, the parameter **T** is of type **any**, as it should be. The methods *Top()* and *Pop()* produce a fatal index violation error if an attempt is made to exercise either of these methods on an empty stack.

The main driver illustrates the proper way to avoid this problem. Before invoking either of these methods, the stack is tested to see whether it is empty.

Here, the stack was implemented in package **main**. Ordinarily, we would create a package **stack**, separate from the **main** package. We did it this way to keep things simple.

In the next section, we present a **Node** implementation of a generic stack.

5.3 Node Implementation of a Generic Stack

Listing 5-3 presents an alternative implementation of stack.

Listing 5-3. Node implementation of generic stack

```go
package main

import (
    "fmt"
)

type Node[T any] struct {
    value T
    next *Node[T]
}

type Stack[T any] struct {
    first *Node[T]
}

// Methods
func (stack *Stack[T]) Push(item T) {
    newNode := Node[T]{item, nil}
    newNode.next = stack.first
    stack.first = &newNode
}
func (stack *Stack[T]) Top() T {
    return stack.first.value
}

func (stack *Stack[T]) Pop() T {
    result := stack.first.value
```

```go
        stack.first = stack.first.next
        return result
}

func (stack Stack[T]) IsEmpty() bool {
        return stack.first == nil
}

func main() {
        // Create a stack of names
        nameStack := Stack[string]{}
        nameStack.Push("Zachary")
        nameStack.Push("Adolf")

        if !nameStack.IsEmpty() {
            topOfStack := nameStack.Top()
            fmt.Printf("\nTop of stack is %s", topOfStack)
        }
        if !nameStack.IsEmpty() {
            poppedFromStack := nameStack.Pop()
            fmt.Printf("\nValue popped from stack is %s",
                        poppedFromStack)
        }

        if !nameStack.IsEmpty() {
            poppedFromStack := nameStack.Pop()
            fmt.Printf("\nValue popped from stack is %s",
                        poppedFromStack)
        }

        if !nameStack.IsEmpty() {
            poppedFromStack := nameStack.Pop()
            fmt.Printf("\nValue popped from stack is %s",
                        poppedFromStack)
        }
```

```
    if !nameStack.IsEmpty() {
        poppedFromStack := nameStack.Pop()
        fmt.Printf("\nValue popped from stack is %s",
                            poppedFromStack)
    }

    // Create a stack of integers
    intStack := Stack[int]{}
    intStack.Push(5)
    intStack.Push(10)
    intStack.Push(0)

    if !intStack.IsEmpty() {
        top := intStack.Top()
        fmt.Printf("\nValue on top of intStack is %d", top)
    }

    if !intStack.IsEmpty() {
        popFromStack := intStack.Pop()
        fmt.Printf("\nValue popped from intStack is
                        %d", popFromStack)
    }

    if !intStack.IsEmpty() {
        popFromStack := intStack.Pop()
        fmt.Printf("\nValue popped from intStack is
                    %d", popFromStack)
    }

    if !intStack.IsEmpty() {
        popFromStack := intStack.Pop()
        fmt.Printf("\nValue popped from intStack is
                        %d", popFromStack)
    }
}
```

```
/* Output
Top of stack is Adolf
Value popped from stack is Adolf
Value popped from stack is Zachary
Value on top of intStack is 0
Value popped from intStack is 0
Value popped from intStack is 10
Value popped from intStack is 5
*/
```

A generic type **Node** is defined along with a generic type **Stack**.

```
type Node[T any] struct {
    value T
    next *Node[T]
}

type Stack[T any] struct {
    first *Node[T]
}
```

We may visualize the data structure as shown in Figure 5-1.

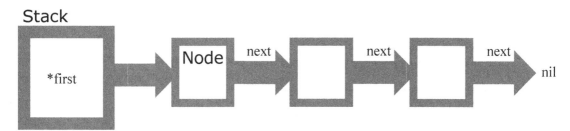

Figure 5-1. *Stack structure*

Function ***main*** is identical to ***main*** in Listing 5-2, and the output is identical. If ***Top()*** or ***Pop()*** are invoked on an empty stack, a memory segment violation would occur. So it is imperative, as in Listing 5-2, to verify that the stack is not empty before invoking either of these methods.

In the next section, we compare the efficiency of the node vs. the slice implementations of stack.

5.4 Compare the Efficiency of Node and Slice Stacks

Which of the two Stack implementations is more efficient?

The slice implementation requires less memory because the node implementation requires the memory overhead of pointers to each succeeding node.

To compare the speed efficiency of these two Stack types, we run a benchmark that pushes 10 million int values onto the stack and then pops the stack until it is empty.

We package the two stack types as **nodestack** and **slicestack** as shown in Listings 5-4 and 5-5. In Listing 5-6, we present the application that compares the speed of these two stack packages.

Listing 5-4. Package nodestack

```
package nodestack

type Node[T any] struct {
    value T
    next *Node[T]
}

type Stack[T any] struct {
    first *Node[T]
}

// Methods
func (stack *Stack[T]) Push(item T) {
    newNode := Node[T]{item, nil}
    // newNode.value = item
    newNode.next = stack.first
    stack.first = &newNode
}

func (stack *Stack[T]) Top() T {
    return stack.first.value
}
```

```go
func (stack *Stack[T]) Pop() T {
    result := stack.first.value
    stack.first = stack.first.next
    return result
}

func (stack Stack[T]) IsEmpty() bool {
    return stack.first == nil
}
```

Listing 5-5. Package slicestack

```go
package slicestack

type Stack[T any] struct {
    items []T
}
// Methods
func (stack *Stack[T]) Push(item T) {
    // item is added to the right-most position in the
    // slice
    stack.items = append(stack.items, item)
}

func (stack *Stack[T]) Pop() T {
    length := len(stack.items)
    returnValue := stack.items[length - 1]
    stack.items = stack.items[:(length - 1)]
    return returnValue
}

func (stack Stack[T]) Top() T {
    length := len(stack.items)
    return stack.items[length - 1]
}

func (stack Stack[T]) IsEmpty() bool {
    return len(stack.items) == 0
}
```

154

Listing 5-6. Speed comparison of nodestack and slicestack

```go
package main

import (
    "example.com/nodestack"
    "example.com/slicestack"
    "time"
    "fmt"
)

const size = 10_000_000

func main() {
    nodeStack := nodestack.Stack[int]{}
    sliceStack := slicestack.Stack[int]{}

    // Benchmark nodeStack
    start := time.Now()
    for i := 0; i < size; i++ {
        nodeStack.Push(i)
    }
    elapsed := time.Since(start)
    fmt.Println("\nTime for 10 million Push() operations on nodeStack: ",
    elapsed)

    start = time.Now()
    for i := 0; i < size; i++ {
        nodeStack.Pop()
    }
    elapsed = time.Since(start)
    fmt.Println("\nTime for 10 million Pop() operations on nodeStack: ",
    elapsed)

    // Benchmark sliceStack
    start = time.Now()
    for i := 0; i < size; i++ {
        sliceStack.Push(i)
    }
```

```
    elapsed = time.Since(start)
    fmt.Println("\nTime for 10 million Push()
                    operations on sliceStack: ", elapsed)

    start = time.Now()
    for i := 0; i < size; i++ {
        sliceStack.Pop()
    }
    elapsed = time.Since(start)
    fmt.Println("\nTime for 10 million Pop() operations
                on sliceStack: ", elapsed)
}
/* Output
Time for 10 million Push() operations on nodeStack:  616.365084ms

Time for 10 million Pop() operations on nodeStack:  29.104829ms

Time for 10 million Push() operations on sliceStack:  148.623915ms

Time for 10 million Pop() operations on sliceStack:  11.485335ms
*/
```

The **slicestack** is significantly faster than the **nodestack**. As always, benchmark results are affected by the processor, the amount of RAM, clock speed, and other factors that vary from machine to machine.

In the next section, we present an application of the stack.

5.5 Stack Application: Function Evaluation

We wish to build a function that takes as input a string representing a mathematical expression with operand symbols from a to z and operators from the set "+", "-", "*", "/", "(", ")".

For example, the input to the function might be "(a + (b - c) / (d * e)". After assigning each operand value a float number, the function must evaluate the expression.

As we will soon see, the stack plays a critical role in designing and implementing this application although this is not at all obvious.

Postfix Evaluation

If one were to perform this computation on a Hewlett-Packard (HP) calculator, the sequence of steps would be the following:

1. Enter the quantity a.

2. Enter the quantity b.

3. Enter the quantity c.

4. Push the subtract button.

5. Enter the quantity d.

6. Enter the quantity e.

7. Push the multiply button.

8. Push the divide button.

9. Push the add button.

Symbolically, this sequence of operations could be written as follows: **abc-de*/+.**

There are no parentheses in the preceding expression. The precedence of operations is encapsulated in the expression. We call the expression a **postfix** representation of the original expression.

To clarify further, suppose **a** were assigned the value 2, **b** the value 3, **c** the value 1, **d** the value 5, and **e** the value 2; the postfix evaluation would be performed as follows:

The operator – (the fourth character in the infix expression) would operate on the previous two operands, b and c. That would produce b – c, which is 2. The next operator, *, would operate on its previous two operands producing d * e, which is 10. The next operator, /, would divide its two previous operands, which are 2 and 10, to produce 0.2. Finally, the last operator, +, would add its two previous operands, which are a and 0.2, producing the answer 2.2.

Following this approach to expression evaluation, we divide the problem into two subproblems. The first subproblem is converting the input expression into a postfix expression. The second subproblem is evaluating this postfix expression.

Each of these subproblems utilizes a stack to accomplish their work.

Function *infixpostfix* in Listing 5-7 converts the infix expression to a postfix form.

Listing 5-7. Conversion from infix to postfix

```go
package main

import (
    "fmt"
    "example.com/nodestack"
)

func precedence(symbol1, symbol2 string) bool {
    // Returns true if symbol1 has a higher precedence
    // than symbol2
    if (symbol1 == "+" || symbol1 == "-") && (symbol2
            == "(" || symbol2 == "/") {
        return false
    } else if (symbol1 == "(" && symbol2 != ")") ||
                symbol2 == "(" {
        return false
    } else {
        return true
    }
}

func isPresent(symbol string, operators []string) bool {
    for i := 0; i < len(operators); i++ {
        if symbol == string(operators[i]) {
            return true
        }
    }
    return false
}

func infixpostfix(infix string) (postfix string) {
    operators := []string{"+", "-", "*", "/", "(", ")"}
    postfix = ""
    nodeStack := nodestack.Stack[string]{}
```

```
for index := 0; index < len(infix); index++ {
    newSymbol := string(infix[index])
    if newSymbol == " " || newSymbol == "\n" {
        continue
    }
    if newSymbol >= "a" && newSymbol <= "z" {
        postfix += newSymbol
    }
    if isPresent(newSymbol, operators) {
        if !nodeStack.IsEmpty() {
            topSymbol := nodeStack.Top()
            if precedence(topSymbol, newSymbol) ==
                true {
                if topSymbol != "(" {
                    postfix += topSymbol
                }
                nodeStack.Pop()
            }
        }
        if newSymbol != ")" {
            nodeStack.Push(newSymbol)
        } else { // Pop nodeStack down to first
                 // left parenthesis
            for {
                if nodeStack.IsEmpty() == true {
                    break
                }
                ch := nodeStack.Top()
                if ch != "(" {
                    postfix += ch
                    nodeStack.Pop()
                } else {
                    nodeStack.Pop()
                    break
                }
```

```
                }
            }
        }
    }
    for {
        if nodeStack.IsEmpty() == true {
            break
        }
        if nodeStack.Top() != "(" {
            postfix += nodeStack.Top()
            nodeStack.Pop()
        }
    }
    return postfix
}

func main() {
    postfix := infixpostfix("a + (b - c) / (d * e)")
    fmt.Println(postfix)
}
// Output: abc-de*/+
```

The ***nodeStack*** is the centerpiece of this algorithm.

We Walk Through Algorithm

Let us "walk" through function ***infixpostfix*** for the infix expression given. We depict the stack with the top of the stack shown on the right and previous items pushed on the stack shown from right to left. The oldest item pushed on our stack depiction is the leftmost item, and the most recent item pushed on our stack is the rightmost item.

The infix expression is "a + (b -c) / (d * e)".

We initialize the ***operators*** slice, the output ***postfix*** string, and the ***nodeStack*** as follows:

```
operators := []string{"+", "-", "*", "/", "(", ")"}
postfix = ""
nodeStack := nodestack.Stack[string]{}
```

160

In a loop that captures each **newSymbol** of the infix expression, if the **newSymbol** is whitespace, we skip the rest of the loop and continue back to the top of the loop.

The first nonwhitespace character is the operand "a". The first "if" statement appends this operand to the **postfix** string.

The next nonwhitespace character is "+". This operator gets pushed onto the **nodeStack**. The state of the system is

```
Stack: +
postfix: a
```

The next nonwhitespace character is "(". Using the **precedence** function and comparing **topSymbol** ("+") with **newSymbol** ("("), it returns false. We therefore push the "(" onto the stack yielding a system state:

```
Stack (top on the right): +  (
postfix: a
```

The next nonwhitespace character gets appended to postfix. The system state is

```
Stack: +  (
postfix: ab
```

We next process the "-" operator. The precedence of "(" with "-" is false, so the "-" operator is pushed onto the stack.

```
Stack (top on the right): +  (  -
postfix: ab
```

Next, we process the operand "c". It gets appended to the result.

```
Stack (top on the right): +  (  -
postfix: abc
```

The next character we process is ")". The precedence is false between "-" and ")". The conditional logic drops us to the "else" clause. As the comment suggests, this causes us to deposit all operands on the stack onto postfix until we encounter the "(" symbol.

```
Stack (top on the right): +
postfix: abc-
```

The next character, "/", gets pushed onto the stack because of the false precedence between "+" and "/".

```
Stack (top on the right): +  /
postfix: abc-
```

For the same reason as earlier, the next symbol, "(", gets pushed onto the stack.

```
Stack (top on the right): +  /  (
postfix: abc-
```

As before, the next operator symbol "*" gets pushed onto the stack.

```
Stack (top on the right): +  /  (  *
postfix: abc-
```

At each stage of this process, the operators on the stack are in increasing order of precedence going from left to right. This is what assures us that the result requires no parentheses.

The next symbol, ")", causes the stack to be cleared up to the "(".

```
Stack (top on the right): +  /
postfix: abc-*
```

With all the symbols from the infix expression processed, only the final loop remains. In this loop, all remaining operator symbols are appended to *postfix* as the stack is popped.

The final state of the system is

```
Stack (top on the right):
postfix: abc-*/+
```

Evaluating Postfix Expression

Next, we grapple with the second part of this problem: evaluating the postfix expression when each operand is assigned a **float64** value.

Listing 5-8 presents the function *evaluate*, which takes as input a postfix expression as well as a map of numeric values for each operand symbol.

Listing 5-8. Evaluating postfix expression

```go
package main

// Snip from Listing 5.7

var values map[string]float64

func evaluate(postfix string) float64 {
    operandStack := nodestack.Stack[float64]{}
    for index := 0; index < len(postfix); index++ {
        ch := string(postfix[index])
        if ch >= "a" && ch <= "z" {
            operandStack.Push(values[ch])
        } else { // ch is an operator
            operand1 := operandStack.Pop()
            operand2 := operandStack.Pop()
            if ch == "+" {
                operandStack.Push(operand1 + operand2)
            } else if ch == "-" {
                operandStack.Push(operand2 - operand1)
            } else if ch == "*" {
                operandStack.Push(operand1 * operand2)
            } else if ch == "/" {
                operandStack.Push(operand2 / operand1)
            }
        }
    }
    return operandStack.Top()
}

func main() {
    postfix := infixpostfix("a + (b - c) / (d * e)")
    fmt.Println(postfix)
    values = make(map[string]float64)
    values["a"] = 10
    values["b"] = 5
    values["c"] = 2
```

```
    values["d"] = 4
    values["e"] = 3
    result := evaluate(postfix)
    fmt.Println("function evaluates to: ", result)
}
// Output: abc-de*/+
// function evaluates to: 10.25
```

In function *evaluate*, another stack, *operandStack*, is the centerpiece. This function is much simpler than the *infixpostfix* function and is left to the reader to walk through a simple example.

The benefit of genericity should be evident. The *nodeStack* in Listing 5-7 used **string** as its type instance, and the *operandStack* in Listing 5-8 used **float64** as its type instance.

In the next section, we consider another application of stack – converting a decimal number to binary.

5.6 Converting Decimal Number to Binary

A much simpler application of stacks is converting a decimal number to binary.

Listing 5-9 shows how to do this.

Listing 5-9. Converting decimal number to binary using a stack

package main

```
import (
    "fmt"
    "example.com/slicestack"
)
```

func convertToBinary(input int) (binary []int) {
```
    binaryNumberStack := slicestack.Stack[int]{}
    for {
        binaryNumberStack.Push(input % 2)
        input = input / 2
```

```
        if input == 0 {
            break
        }
    }
    binary = []int{}
    for {
        if !binaryNumberStack.IsEmpty() {
            binary = append(binary,
                    binaryNumberStack.Pop())
        } else {
            break
        }
    }
    return binary
}

func main() {
    number := 1_000_000
    binaryNumber := convertToBinary(number)
    fmt.Printf("\n%d converted to binary is \n%v",
                        number, binaryNumber)
}
/* Output
1000000 converted to binary is
[1 1 1 1 0 1 0 0 0 0 1 0 0 1 0 0 0 0 0 0]
*/
```

Here, the ***binaryNumberStack[int]*** is used to reverse the sequence of 0's and 1's produced by finding the sequence of remainders, ***input % 2***, as input is reduced by a factor of 2 at every iteration.

In the next section, we present another application of stack, finding a path through a maze.

5.7 Maze Application

Although the study of data structures, like the stack and many others to be explored later, is interesting, it is when data structures and their associated operations are deployed in applications that they come to life.

In this section, we present a more complex application in which the stack plays a central role.

Note   This application is an adaptation of an example presented in Section 3.2 of *Data Structures Using Modula-2* by Richard Sincovec and Richard Wiener (John Wiley, 1986) and later implemented in *C# in Modern Software Development Using C#.Net* by Richard Wiener (Thompson Learinng, 2007).

We represent a maze with a two-dimensional list of 0's and 1's. Cells with value 1 represent obstacles that block a maze path. Cells with value 0 represent possible maze path locations. Given such a matrix file of 0's and 1's and given the starting location and ending location, the goal is to write a Go application that finds a path from starting location to ending location, if one or more such paths exist.

We wish to avoid a brute-force strategy that enumerates every possible path from starting point to ending point.

Efficient Strategy for Maze Path Using a Stack

Using a **stack**, we can develop an efficient strategy, which is outlined as follows.

At any location along the maze path, the next move can be chosen from among the eight adjacent locations (north, northeast, east, southeast, south, southwest, west, and northwest) providing that the given location has value 0 (is open). We let the program make a random choice among the open adjacent locations. The program will possibly produce different viable paths each time it is run.

Since a path cannot visit the same location more than once, we set the value along each cell in the path from 0 to 1.

After each move, we push the current position along with the direction of the move to be made onto the path stack. If the path hits a dead end as many typically will, we can backtrack and access the last safe position and continue from there.

166

More formally, our maze algorithm uses a stack as its central control mechanism and is the following:

1. Load the maze file, number of rows, number of columns, and the starting and ending locations.

2. Initialize a path stack that holds path objects. We use a generic stack with the base type **T** of type **Path**.

3. A **path** object contains a coordinate within the maze, a current move direction, and a list of available move directions.

4. Choose an initial move direction from among the open neighboring locations.

5. As each move direction is attempted, delete it from the list of eight possible move directions.

6. Construct a new path object from the starting point, an initial move direction, and a list of remaining move directions.

7. Push the initial path object onto the stack.

8. While the stack is not empty, get the path object at the top of the stack by popping the stack.

9. Start a loop: While the current path object has more available moves, choose one of the available locations randomly and set its value from 0 to 1. Construct a new path object and push it onto the stack. While the stack is not empty, get the path object at the top of the stack by popping the stack.

Building Infrastructure for Maze Application

Before we plunge into the maze implementation, we build some infrastructure by defining some relevant types and their operations – some abstract data types.

Listing 5-10 introduces the basic types needed for the maze application.

Listing 5-10. Type infrastructure for maze application

package main

```
import (
    "fmt"
    "math/rand"
    "time"
)
```

```
// Direction abstraction
```
type Direction int

const (
**   N int = 0**
**   NE = 1**
**   E = 2**
**   SE = 3**
**   S = 4**
**   SW = 5**
**   W = 6**
**   NW = 7**
**   NotAvailable = 8**
)

func (d Direction) String() string {
```
    switch d {
    case 0:
        return "north"
    case NE:
        return "north-east"
    case E:
        return "east"
    case SE:
        return "south-east"
    case S:
        return "south"
```

```go
    case SW:
        return "south-west"
    case W:
        return "west"
    case NW:
        return "north-west"
    case NotAvailable:
        return "not available"
    }
    return "unknown"
}

func (d Direction) PrintDirection() {
    fmt.Println("direction: ", d)
}

// Point abstraction
type Point struct {
    x, y int
}

func (p Point) Equals(other Point) bool {
    return p.x == other.x && p.y == other.y
}

func (p Point) PrintPoint() {
    fmt.Printf("<%d, %d>\n", p.x, p.y)
}

// Path abstraction
type Path struct {
    point Point
    move Direction
    movesAvailable []Direction
}

func NewPath(point Point) Path {
    path := Path{point, Direction(NotAvailable),
                 []Direction{}}
```

```go
        path.move = NotAvailable
        // Initially all directions available
        path.movesAvailable = []Direction{O, NE, E, SE, S, SW, W, NW}
        return path
}

func (path *Path) RandomMove() Direction {
        // Returns value of move and changes the receiver
        indicesAvailable := []int{}
        for index := 0; index < 8; index++ {
                if path.movesAvailable[index] != NotAvailable {
                        indicesAvailable =
                                append(indicesAvailable, index)
                }
        }
        count := len(indicesAvailable)
        if count > 0 {
                randomIndex := rand.Intn(count)
                path.move =
                                path.movesAvailable[indicesAvailable[randomIndex]]
                path.movesAvailable[indicesAvailable[randomIndex]]
                                = NotAvailable
                return path.move
        } else {
                return NotAvailable
        }
}

func main() {
        rand.Seed(time.Now().UnixNano())
        myDirection := Direction(6)
        myDirection.PrintDirection()
        myPoint := Point{3, 4}
        myPoint.PrintPoint()
        result := myPoint.Equals(Point{3, 4})
        fmt.Println(result)
```

```
    myPath := NewPath(Point{3, 4})
    randomMove := myPath.RandomMove()
    fmt.Println(randomMove)
    fmt.Println(myPath)
}
/* Output
direction:  west
<3, 4>
true
south
{{3 4} 4 [0 1 2 3 8 5 6 7]}
*/
```

The method *RandomMove* changes the receiver and returns the direction of the move.

Go does not support enum types, so we simulate an enum type by defining

type Direction int

Creating this new type allows us to protect entities of this type from being manipulated and possibly corrupted like they were ordinary integers.

We define a set of constants representing the nine directions that are possible (if we consider *NotAvailable* to be one of these).

Function *main* does nothing useful but is there to illustrate how variables of each type can be created and used.

Now we are ready to introduce the **Maze** abstraction and write this application. Since we will need a stack (we will use a **slicestack** for this application although a **nodestack** would do just as well), we will create a separate subdirectory for the **Maze** functionality (the code in package **main**) and import the **slicestack**. We will create a **go.mod** file in the subdirectory **mainmaze** that contains the **main** package. The go.mod file is

module example.com/main

go 1.18

replace example.com/slicestack => ../slicestack

require example.com/slicestack v0.0.0-00010101000000-000000000000

171

The **Maze** type is defined as follows:

```
type Maze struct {
    rows, cols int
    start, end Point
    mazefile   string
    barriers   [][]bool
    current    Path
    moveCount  int
    pathStack  slicestack.Stack[Path]
    gameOver   bool
}
```

The field *barriers*, which defines the locations that are either blocked or open, is a two-dimensional slice of bool. A rune of "1" in the *mazefile* indicates a blocked location, and a rune of "0" indicates an open location.

The field *pathStack* is a *slicestack.Stack* with *Path* as its generic type.

The function *NewMaze* creates an instance of **Maze** as follows:

```
func NewMaze(rows int, cols int, start Point, end
                Point, mazefile string) (maze Maze) {
    maze.rows = rows
    maze.cols = cols
    maze.start = start
    maze.end = end

    // Initialize maze.barriers
    maze.barriers = make([][]bool, rows)
    for i := range maze.barriers {
        maze.barriers[i] = make([]bool, cols)
    }

    file, err := os.Open(mazefile)
    if err != nil {
        log.Fatal(err)
    }
    scanner := bufio.NewScanner(file)
    scanner.Split(bufio.ScanLines)
```

```
var textlines []string
for scanner.Scan() {
    textlines = append(textlines, scanner.Text())
}
defer file.Close()
for row := 0; row < rows; row++ {
    line := textlines[row]
    for col := 0; col < cols; col++ {
        if string(line[col]) == "1" {
            maze.barriers[row][col] = true
        } else {
            maze.barriers[row][col] = false
        }
    }
}
maze.current = NewPath(start)
maze.pathStack = slicestack.Stack[Path]{}
maze.pathStack.Push(maze.current)
maze.barriers[start.x][start.y] = true
return maze
```
}

The two-dimensional slice *barriers* are initialized by allocating storage for the given number of rows and then for each row allocating storage for the columns.

The input text file, *mazefile*, is read line by line using *NewScanner* from package **bufio**.

Then the barriers slice is assigned true at a given row and column if a "1" is present and false if a "0" is present.

A support function, *NewPosition*, returns a *Point* based on the *oldPosition* and the *move* direction and is given as follows:

```
func NewPosition(oldPosition Point, move Direction)
                Point {
    if move == Direction(N) {
        return Point{oldPosition.x, oldPosition.y - 1}
    } else if move == NE {
        return Point{oldPosition.x + 1, oldPosition.y - 1}
```

```
    } else if move == E {
        return Point{oldPosition.x + 1, oldPosition.y}
    } else if move == SE {
        return Point{oldPosition.x + 1, oldPosition.y + 1}
    } else if move == S {
        return Point{oldPosition.x, oldPosition.y + 1}
    } else if move == SW {
        return Point{oldPosition.x - 1, oldPosition.y + 1}
    } else if move == W {
        return Point{oldPosition.x - 1, oldPosition.y}
    } else {
        return Point{oldPosition.x - 1, oldPosition.y - 1}
    }
}
```

The main program logic for advancing through the maze is given in method **StepAhead**. This function returns a new position and backtracks location, each of type **Point**.

This function is given as follows:

```
func (m *Maze) StepAhead() (Point, Point) {
    validMove := false
    backTrackPoint := None
    newPos := None
    for {
        if m.gameOver || validMove ||
                m.pathStack.IsEmpty() {
            break
        }
        validMove = false
        m.current = m.pathStack.Pop()
        m.moveCount += 1
        nextMove := m.current.RandomMove()
        for {
            if validMove || nextMove == NotAvailable {
                break
            }
```

```
        newPos = NewPosition(m.current.point,
                        m.current.move)
        if m.barriers[newPos.y][newPos.x] == false
        {
            validMove = true
            if newPos.Equals(m.end) {
                for {
                    if m.pathStack.IsEmpty() ==
                            true {
                        break
                    }
                    m.pathStack.Pop()
                }
                m.gameOver = true
            }
            m.barriers[newPos.y][newPos.x] = true
            m.pathStack.Push(m.current)
            newPathObject := NewPath(newPos)
            m.pathStack.Push(newPathObject)
        } else {
            nextMove = m.current.RandomMove()
        }
    }
    if !validMove && !m.pathStack.IsEmpty() {
        fmt.Printf("\nBacktrack from %v to %v\n",
                m.current.point,
                    m.pathStack.Top().point)
        backTrackPoint = m.pathStack.Top().point
    }
}
if m.pathStack.IsEmpty() {
    fmt.Println("No solution is possible")
    return None, None
}
return newPos, backTrackPoint
}
```

Two nested for-loops control the logic of finding the next position in the maze. The outer loop terminates if the ***gameOver*** field of the maze ***m*** is true or if a ***validMove*** is true or if the ***pathStack*** of the maze is empty. If the ***pathStack*** is empty, then the application terminates with the message "No solution is possible." The inner loop terminates if a valid move is found, or the next random move is **NotAvalable**.

The pieces of this application fit together tightly and are moderately complex. It should be evident that the ***slicestack.Stack[Path]*** plays a central role in moving through the maze.

Completed Maze App

Listing 5-11 presents the complete maze app with a ***main*** driver and output from a typical run. The maze file, **maze.txt**, for this run is illustrated in Figure 5-2.

```
1111111111111111111111111111111111111111
1011110111111111111111111111111111111111
1101101111111111111111111111111111111111
1000011111111111111110111111100001111111
1111011111111111111110111111111111111111
1111101111111111111110111111111111111111
1111110111111111111110111111111111111111
1111101011111111111110111111111111111111
1111011100011111111110111111111111111111
1111011111101111111110111111111111111111
1111011111101111111110111111111111111111
1111011111101111111110111111111111111111
1111011111110000011101111111111111111111
1111101111111111000011111111111111111111
1111100111111111110111111111111111111111
1111110011111111110111111111111111111111
1111101111111111110111111111111111111111
1111011111111111111000001110000000000111
1111101111111111111111000000111111111111
```

Figure 5-2. *maze.txt*

```
1111111011111111111111111111111111111111
1100101011111111111111111111111111111111
1111001001111111111111111111111111111111
1111111100000000000000000000000011111111
1111111111111111111111111111110101111111
1111111111111111111111111111110110111111
1111111111111111111111111111110111011111
1111111111111111111111111111110111101111
1111111111111111111111111111110111110111
1111111111111111111111111111110111110111
1111111111111111111111111111101111110111
1111111111111111111111111111011111110111
1111111111111111111111111110111111110111
1111111111111111111111111101111111110111
1111111111111111111111111011111111110111
1111111111111111111111111111111111110111
1111111111111111111111111111111111110111
1111111111111111111111111111111111110111
1111111111111111111111111111111111110111
1111111111111111111111111111111111111001
1111111111111111111111111111111111111111
```

Figure 5.2. (*continued*)

You can see from the sequence of zeros that a solution is possible if the starting location is <1, 1> and the ending location is <38, 38> in this 40 × 40 grid of possible locations. You can also see the possibility of several side tracks that lead to dead ends.

Listing 5-11. Maze application

```
// MAZE application
package main

import (
    "bufio"
    "example.com/slicestack"
    "fmt"
    "log"
```

```
    "math/rand"
    "os"
    "time"
)

// Snip from Listing 5.10

// *******************************
// MAZE abstraction
type Maze struct {
    rows, cols int
    start, end Point
    mazefile   string
    barriers   [][]bool
    current    Path
    moveCount  int
    pathStack  slicestack.Stack[Path]
    gameOver   bool
}

func NewMaze(rows int, cols int, start Point, end
            Point, mazefile string) (maze Maze) {
    maze.rows = rows
    maze.cols = cols
    maze.start = start
    maze.end = end

    // Initialize maze.barriers
    maze.barriers = make([][]bool, rows)
    for i := range maze.barriers {
        maze.barriers[i] = make([]bool, cols)
    }

    file, err := os.Open(mazefile)
    if err != nil {
        log.Fatal(err)
    }
```

```
scanner := bufio.NewScanner(file)
scanner.Split(bufio.ScanLines)
var textlines []string
for scanner.Scan() {
    textlines = append(textlines, scanner.Text())
}
defer file.Close()
for row := 0; row < rows; row++ {
    line := textlines[row]
    for col := 0; col < cols; col++ {
        if string(line[col]) == "1" {
            maze.barriers[row][col] = true
        } else {
            maze.barriers[row][col] = false
        }
    }
}
maze.current = NewPath(start)
maze.pathStack = slicestack.Stack[Path]{} // generic instance
maze.pathStack.Push(maze.current)
maze.barriers[start.x][start.y] = true
return maze
}

func NewPosition(oldPosition Point, move Direction)
                Point {
    if move == Direction(N) {
        return Point{oldPosition.x, oldPosition.y - 1}
    } else if move == NE {
        return Point{oldPosition.x + 1, oldPosition.y - 1}
    } else if move == E {
        return Point{oldPosition.x + 1, oldPosition.y}
    } else if move == SE {
        return Point{oldPosition.x + 1, oldPosition.y + 1}
    } else if move == S {
        return Point{oldPosition.x, oldPosition.y + 1}
```

```go
    } else if move == SW {
        return Point{oldPosition.x - 1, oldPosition.y + 1}
    } else if move == W {
        return Point{oldPosition.x - 1, oldPosition.y}
    } else {
        return Point{oldPosition.x - 1, oldPosition.y - 1}
    }
}

func (m *Maze) StepAhead() (Point, Point) {
    validMove := false
    backTrackPoint := None
    newPos := None
    for {
        if m.gameOver || validMove ||
                    m.pathStack.IsEmpty() {
            break
        }
        validMove = false
        m.current = m.pathStack.Pop()
        m.moveCount += 1
        nextMove := m.current.RandomMove()
        for {
            if validMove || nextMove == NotAvailable {
                break
            }
            newPos = NewPosition(m.current.point, m.current.move)
            if m.barriers[newPos.y][newPos.x] == false
            {
                validMove = true
                if newPos.Equals(m.end) {
                    for {
                        if m.pathStack.IsEmpty() ==
                                true {
                            break
                        }
```

```
                    m.pathStack.Pop()
                }
                m.gameOver = true
            }
            m.barriers[newPos.y][newPos.x] = true
            m.pathStack.Push(m.current)
            newPathObject := NewPath(newPos)
            m.pathStack.Push(newPathObject)
        } else {
            nextMove = m.current.RandomMove()
        }
    }
    if !validMove && !m.pathStack.IsEmpty() {
        fmt.Printf("\nBacktrack from %v to %v\n",
                m.current.point,
                m.pathStack.Top().point)
        backTrackPoint = m.pathStack.Top().point
    }
}
if m.pathStack.IsEmpty() {
    fmt.Println("No solution is possible")
    return None, None
}
return newPos, backTrackPoint
}

// *******************************************

func main() {
    rand.Seed(time.Now().UnixNano())
    start := Point{1, 1}
    end := Point{38, 38}
    maze := NewMaze(40, 40, start, end, "maze.txt")
    newPos, _ := maze.StepAhead()
    time.Sleep(1 * time.Second)
```

```
    if newPos != None {
        fmt.Println(newPos)
    }
    for {
        if newPos == None || newPos.Equals(end) {
            break
        }
        newPos, _ = maze.StepAhead()
        time.Sleep(100 * time.Millisecond)
        if newPos != None {
            fmt.Println(newPos)
        }
    }
    if newPos.Equals(end) {
        fmt.Println("SUCCESS!  Reached ", end)
    }
}
/* Output
{2 2}
{1 3}
{2 3}
{3 3}
{4 4}
{4 3}
{5 2}
{6 1}

Backtrack from {6 1} to {5 2}

Backtrack from {5 2} to {4 3}

Backtrack from {4 3} to {4 4}
{5 5}
{6 6}
{5 7}
{4 8}
{4 9}
```

{4 10}
{4 11}
{4 12}
{5 13}
{6 14}
{6 15}
{5 16}
{4 17}
{5 18}
{6 19}
{5 20}
{5 21}
{4 21}
{3 20}
{2 20}
Backtrack from {2 20} to {3 20}

Backtrack from {3 20} to {4 21}

Backtrack from {4 21} to {5 21}

Backtrack from {5 21} to {5 20}

Backtrack from {5 20} to {6 19}
{7 20}
{8 21}
{8 22}
{7 21}

Backtrack from {7 21} to {8 22}
{9 22}
{10 22}
{11 22}
{12 22}
{13 22}
{14 22}
{15 22}
{16 22}

```
{17 22}
{18 22}
{19 22}
{20 22}
{21 22}
{22 22}
{23 22}
{24 22}
{25 22}
{26 22}
{27 22}
{28 22}
{29 22}
{30 23}
{31 22}
{32 23}
{33 24}
{34 25}
{35 26}
{36 27}
{36 28}
{36 29}
{36 30}
{36 31}
{36 32}
{36 33}
{36 34}
{36 35}
{36 36}
{36 37}
{37 38}
{38 38}
SUCCESS!   Reached   {38 38}
*/
```

In the output shown, there were three dead end detours. The ***pathStack*** enabled backtracking recovery from each of these detours and the eventual successful path through the maze.

5.8 Summary

In this chapter, we showed two implementations of a generic stack. We then proceeded with several applications of stack including algebraic function evaluation, converting decimal to binary and finding the path through a maze.

In the next chapter, we focus on the queue and list data structures.

CHAPTER 6

Queues and Lists

Queue is another relatively simple data type. It has many practical uses in application development.

A queue organizes data in a first-in, first-out (FIFO) manner. Because of FIFO, the most obvious application is to model a waiting line. This could be a line of customers waiting for some service, a print job waiting in a print queue, a concurrent process waiting for CPU access, and many other applications that require waiting lines. New items are inserted into the back of a queue, and items are removed from the front of the queue. The queue maintains the order in which the items are inserted.

We present two implementations of **Queue** in this chapter and compare their efficiency. We also present several applications of **Queue**.

Deque is more general than a **Queue**. It allows insertion and deletion from the front as well as the back of the structure. We present an implementation of **Deque** and an application that uses **Deque**.

PriorityQueue is a specialized type of **Queue**. We show an implementation of *PriorityQueue* and an application involving airline passengers.

List is a more general data type than a **Queue.** Items can be inserted in the front, back, or anywhere in the middle. We present the implementation of a singly linked and doubly linked list.

In the next section, we define the Queue abstract data type (ADT).

© Richard Wiener 2022
R. Wiener, *Generic Data Structures and Algorithms in Go*, https://doi.org/10.1007/978-1-4842-8191-8_6

6.1 Queue ADT

There are six operations that characterize a Queue ADT.

Insert(item) – Adds item to the queue

Remove( ) item – Removes and returns the first item inserted in the queue

First( ) item – Accesses the first item inserted in the queue without altering the queue

Size int – Returns the number of items in the queue

Range( ) – Returns an Iterator

Empty( ) – Returns a bool, true if the Iterator it is applied to has no items

Next( ) – Returns the next item in the Iterator

We present two implementations of **Queue**: slice based and node based. In the next section, we focus on a slice-based implementation of Queue.

6.2 Implementation of Slice Queue

Listing 6-1 presents a generic slice implementation of **Queue** in package *slicequeue*.

The *Queue* struct contains a field, *items*, a slice of generic type *T*.

Listing 6-1. Generic slice implementation of Queue

```
package slicequeue

type Queue[T any] struct {
    items []T
}

type Iterator[T any] struct {
    next int // index in items
    items []T
}
```

```go
// Queue Methods
func (queue *Queue[T]) Insert(item T) {
    // item is added to the right-most position in the slice
    queue.items = append(queue.items, item)
}

func (queue *Queue[T]) Remove() T {
    returnValue := queue.items[0]
    queue.items = queue.items[1:]
    return returnValue
}

func (queue Queue[T]) First() T {
    return queue.items[0]
}

func (queue Queue[T]) Size() int {
    return len(queue.items)
}

func (queue *Queue[T]) Range() Iterator[T] {
    return Iterator[T]{0, queue.items}
}

// Iterator Methods
func (iterator *Iterator[T]) Empty() bool {
    return iterator.next == len(iterator.items)
}

func (iterator *Iterator[T]) Next() T {
    returnValue := iterator.items[iterator.next]
    iterator.next++
    return returnValue
}
```

The FIFO protocol of **Queue** is achieved by inserting new items in the rightmost position of the *items* slice and removing items from the leftmost position, index 0, in the *items* slice.

Iterator

An *Iterator* type is a struct containing an index *next* and the *items* slice.

The *Empty* method on *Iterator* is true if the iterator field *next* equals the length of the *items* slice.

The *Next* method on *Iterator* returns the value *T* in index *next* of the *items* slice.

The *Range* method on **Queue** returns an **Iterator.**

Listing 6-2 shows a simple main driver program that exercises a generic queue.

Listing 6-2. Driver Program for Generic Queue

```go
package main
import (
    "fmt"
    "example.com/slicequeue"
)

func main() {
    myQueue := slicequeue.Queue[int]{}
    myQueue.Insert(15)
    myQueue.Insert(20)
    myQueue.Insert(30)
    myQueue.Remove()
    fmt.Println(myQueue.First())
    queue := slicequeue.Queue[float64]{}
    for i := 0; i < 10; i++ {
        queue.Insert(float64(i))
    }
    iterator := queue.Range()
    for {
        if iterator.Empty() {
            break
        }
        fmt.Println(iterator.Next())
    }
    fmt.Println("queue.First() = ", queue.First())
}
```

```
/* Output
20
0
1
2
3
4
5
6
7
8
9
queue.First() =  0
*/
```

The package *slicequeue* is imported. A queue with type **int** and another queue with type **float64** are defined and exercised. It is noted that when the **float64** queue is constructed and the values are displayed using an iterator, the state of the queue is not changed.

In the next section, we present the implementation of a node-based **Queue**.

6.3 Implementation of Node Queue

Listing 6-3 presents a node implementation of **Queue**.

Listing 6-3. Generic node implementation of queue

```
package nodequeue

type Node[T any] struct {
    item T
    next *Node[T]
}
```

```
type Queue[T any] struct {
    first, last *Node[T]
    length int
}

type Iterator[T any] struct {
    next *Node[T]
}

// Methods
func (queue *Queue[T]) Insert(item T) {
    newNode := &Node[T]{item, nil}
    if queue.first == nil {
        queue.first = newNode
        queue.last = queue.first
    } else {
        queue.last.next = newNode
        queue.last = newNode
    }
    queue.length +=1
}

func (queue *Queue[T]) Remove() T {
    returnValue := queue.first.item
    queue.first = queue.first.next
    if queue.first == nil {
        queue.last = nil
    }
    return returnValue
}

func (queue Queue[T]) First() T {
    return queue.first.item
}

func (queue Queue[T]) Size() int {
    return queue.length
}
```

```go
func (queue *Queue[T]) Range() Iterator[T] {
    return Iterator[T]{queue.first}
}

func (iterator *Iterator[T]) Empty() bool {
    return iterator.next == nil
}

func (iterator *Iterator[T]) Next() T {
    returnValue := iterator.next.item
    if iterator.next != nil {
        iterator.next = iterator.next.next
    }
    return returnValue
}
```

A generic **Node** type is defined containing an **item** field of type **T** and a **next** field, a pointer to **Node**. This recursive structure is similar to what we did in defining a node in **nodestack**.

The **Queue** type is a struct containing two pointers to **Node**, **first** and **last**. They point to the beginning and end of the queue.

The **Insert** method creates a **first** value if the queue is empty and sets **last** to equal **first**. If the queue already has a non-nil **first** value, it links the current **last** value to the new node and replaces **last** with a pointer to this new node. The **first** value is unaffected.

The **Remove** method returns the **item** in the **first** Node and resets **first** to its **first.next** link. If **first** becomes nil, then the **last** field is also set to nil; otherwise, it is unaffected.

The **Iterator** is a struct with a **next** field that points to a **Node**.

The **Range** method returns an **Iterator** that contains a **next** field pointing to the **first** item in the queue.

The **Empty** method on **Iterator** returns true if the **next** field points to nil; otherwise, it returns false.

Finally, the **Next** method on **Iterator** returns the value in the **next** Node and advances the field **iterator.next** to the **iterator.next.next** link.

A main driver program that exercises the **queuenode** is the same as in Listing 6-2 except the package "example.com/nodequeue" is used.

In the next section, we compare the performance of a slice-based **Queue** with a node-based **Queue**.

6.4 Comparing the Performance of Slice and Node Queue

Listing 6-4 presents a program that compares the execution time of inserting and removing items from a *slicequeue* and a *nodequeue*.

Listing 6-4. Benchmarking the performance of slicequeue and nodequeue

```go
// We compare the performance of slicequeue and nodequeue
package main

import (
    "fmt"
    "example.com/nodequeue"
    "example.com/slicequeue"
    "time"
)

const size = 1_000_000

func main() {
    sliceQueue := slicequeue.Queue[int]{}
    nodeQueue := nodequeue.Queue[int]{}
    start := time.Now()
    for i := 0; i < size; i++ {
        sliceQueue.Insert(i)
    }
    elapsed := time.Since(start)
    fmt.Println("Time for inserting 1 million ints in sliceQueue is",
    elapsed)

    start = time.Now()
    for i := 0; i < size; i++ {
        nodeQueue.Insert(i)
    }
```

```
    elapsed = time.Since(start)
    fmt.Println("Time for inserting 1 million ints in nodeQueue is",
    elapsed)

    start = time.Now()
    for i := 0; i < size; i++ {
        sliceQueue.Remove()
    }
    elapsed = time.Since(start)
    fmt.Println("Time for removing 1 million ints from sliceQueue is",
    elapsed)

    start = time.Now()
    for i := 0; i < size; i++ {
        nodeQueue.Remove()
    }
    elapsed = time.Since(start)
    fmt.Println("Time for removing 1 million ints from nodeQueue is",
    elapsed)
}
/* Output
Time for inserting 1 million ints in sliceQueue is 18.841914ms
Time for inserting 1 million ints in nodeQueue is 30.275662ms
Time for removing 1 million ints from sliceQueue is 1.413447ms
Time for removing 1 million ints from nodeQueue is 2.818313ms
*/
```

As expected, the slice queue is significantly faster than the node queue because of the overhead associated with pointer access in the node-based queue.

In the next section, we introduce and implement the **Deque** data structure.

6.5 Deque

A **Deque** is a queue in which items may be inserted or deleted from the front or the back of the structure.

Listing 6-5 presents a slice implementation of a generic **Deque**.

Listing 6-5. Generic slice implementation of Deque

```go
package deque

type Deque[T any] struct {
    items []T
}

func (deque *Deque[T]) InsertFront(item T) {
    deque.items = append(deque.items, item) // Expands deque.items
    for i := len(deque.items) - 1; i > 0 ; i-- {
        deque.items[i] = deque.items[i - 1]
    }
    deque.items[0] = item
}

func (deque *Deque[T]) InsertBack(item T) {
    deque.items = append(deque.items, item)
}

func (deque *Deque[T]) First() T {
    return deque.items[0]
}

func (deque *Deque[T]) RemoveFirst() T {
    returnValue := deque.items[0]
    deque.items = deque.items[1:]
    return returnValue
}

func (deque *Deque[T]) Last() T {
    return deque.items[len(deque.items) - 1]
}

func (deque *Deque[T]) RemoveLast() T {
    length := len(deque.items)
    returnValue := deque.items[length - 1]
    deque.items = deque.items[:(length - 1)]
    return returnValue
}
```

```go
func (deque *Deque[T]) Empty() bool {
    return len(deque.items) == 0
}
```

Listing 6-6 presents a simple driver program that uses **Deque**.

Listing 6-6. Exercising Deque

```go
package main
import (
    "fmt"
    "example.com/deque"
)

func main() {
    myDeque := deque.Deque[int]{}
    myDeque.InsertFront(5)
    myDeque.InsertBack(10)
    myDeque.InsertFront(2)
    myDeque.InsertBack(12) // 2 5 10 12
    fmt.Println("myDeque.First() = ", myDeque.First())
    fmt.Println("myDeque.Last() = ", myDeque.Last())

    myDeque.RemoveLast()
    myDeque.RemoveFirst()
    fmt.Println("myDeque.First() = ", myDeque.First())
    fmt.Println("myDeque.Last() = ", myDeque.Last())
}
/* Output
myDeque.First() =  2
myDeque.Last() =  12
myDeque.First() =  5
myDeque.Last() =  10
*/
```

In the next section, we present an application that uses **Deque**.

6.6 Deque Application

Given an array and an integer **k**, find the maximum value for every contiguous subarray of size k.

As an example, consider the following problem:

Input array: **input := []int{9, 1, 1, 0, 0, 0, 1, 0, 6, 8}** with k = 3

Max of 9, 1, 1 is **9**.

Max of 1, 1, 0 is **1**.

Max of 1, 0, 0 is **1**.

Max of 0, 0, 0 is **0**.

Max of 0, 0, 1 is **1**.

...

So the output is **[9 1 1 0 1 1 6 8]**.

Listing 6-7 presents a simple brute-force solution to this problem.

Listing 6-7. Brute-force solution to the maximum contiguous array problem

```go
package main

import (
    "fmt"
)

func MaxSubarray(input []int, k int) (output []int) {
    for first := 0; first <= len(input) - k; first++ {
        max := input[first]
        for second := 0; second < k; second++ {
            if input[first + second] > max {
                max = input[first + second]
            }
        }
        output = append(output, max)
    }
    return output
}
```

```go
func main() {
    input := []int{3, 1, 6, 4, 2, 10, 5, 9}
    output := MaxSubarray(input, 3)
    fmt.Println("Output = ", output)
}
/* Output
Output =  [6 6 6 10 10 10]
*/
```

Because of the nested loops, the computational complexity of this solution is **O(n * k)**, where **n** is the size of the input slice.

Can we do better? This would be useful if n and k were large. We can do much better using the services of a **Deque**.

Consider function *MaxSubarrayUsingDeque* as follows:

```go
func MaxSubarrayUsingDeque(input []int, k int) (output []int) {

    deque := deque.Deque[int]{}

    var index int
    // First window
    for index = 0; index < k; index++ {
        for {
            if deque.Empty() || input[index] < input[deque.Last()] {
                break
            }
            deque.RemoveLast()
        }
        deque.InsertBack(index)
    }

    for ; index < len(input); index++ {
        output = append(output, input[deque.First()])

        // Remove elements out of the window
        for {
            if deque.Empty() || deque.First() > index - k {
                break
            }
```

199

```
            deque.RemoveFirst()
    }
    // Remove values smaller than the element currently being added
    for {
        if deque.Empty() || input[index] < input[deque.Last()] {
            break
        }
        deque.RemoveLast()
    }
    deque.InsertBack(index)
    }
    output = append(output, input[deque.First()])
    return output
}
```

Let us walk through the function for a portion of the example before.

A *deque* with generic type **int** is initialized to empty.

Since *deque* is empty, we break out of the inner for-loop and insert index 0 into the deque and then advance *index* from 0 to 1.

Since **input[1]** is less than **input[0]**, we again break out of the inner for-loop and insert index 1 into the back of the deque so the deque contains [0 1]. We advance index to 2.

Since **input[2]** is not less than **index[1]**, we remove the last element, 1, from the deque, leaving the deque as [0]. Since **index[2]** is less than **input[0]**, we break out of the inner loop and insert index 2 to the back of the deque producing [0 2]. The outer for-loop is done. We are assured that the first element in the deque is the largest in the deque.

In the second outer for-loop, we append *input[deque,First()]* to the *output*, namely, the value of 9.

The logic of the second outer for-loop mirrors the first outer for-loop. First, the deque is purged of values out of the index window of the deque, which gets shifted by one to the right after each iteration. Then the deque is filled with the next k values, and the values are rotated so that the first value in the deque is largest.

The computational complexity of this algorithm is **O(n)**.

Listing 6-8 compares the performance of the brute-force algorithm with the deque-based algorithm.

Listing 6-8. Comparing the performance of the brute-force algorithm with the deque-based algorithm

```go
package main

import (
    "fmt"
    "example.com/deque"
    "time"
    "math/rand"
)

const size = 1_000_000

func MaxSubarrayBruteForce(input []int, k int) (output []int) {
    for first := 0; first <= len(input) - k; first++ {
        max := input[first]
        for second := 0; second < k; second++ {
            if input[first + second] > max {
                max = input[first + second]
            }
        }
        output = append(output, max)
    }
    return output
}

func MaxSubarrayUsingDeque(input []int, k int) (output []int) {

    deque := deque.Deque[int]{}

    var index int
    // First window
    for index = 0; index < k;  index++ {
        for {
            if deque.Empty() || input[index] < input[deque.Last()] {
                break
            }
        }
```

```go
            deque.RemoveLast()
        }
        deque.InsertBack(index)
    }

    for ; index < len(input); index++ {
        output = append(output, input[deque.First()])

        // Remove elements out of the window
        for {
            if deque.Empty() || deque.First() > index - k {
                break
            }
            deque.RemoveFirst()
        }
        // Remove values smaller than the element currently being added
        for {
            if deque.Empty() || input[index] < input[deque.Last()] {
                break
            }
            deque.RemoveLast()
        }
        deque.InsertBack(index)
    }
    output = append(output, input[deque.First()])
    return output
}

func main() {
    input := []int{9, 1, 1, 0, 0, 0, 1, 0, 6, 8}
    output1 := MaxSubarrayBruteForce(input, 3)
    fmt.Println("Output = ", output1)

    output2 := MaxSubarrayUsingDeque(input, 3)
    fmt.Println("Output = ", output2)

    // Benchmark performance of two algorithms
    input = []int{}
```

```
    for i := 0; i < size; i++ {
        input = append(input, rand.Intn(1000))
    }
    start := time.Now()
    MaxSubarrayUsingDeque(input, 10000)
    elapsed := time.Since(start)
    fmt.Println("Using Deque: ", elapsed)

    start = time.Now()
    MaxSubarrayBruteForce(input, 10000)
    elapsed = time.Since(start)
    fmt.Println("Using Brute Force: ", elapsed)
}
/* Output
Output = [9 1 1 0 1 1 6 8]
Output = [9 1 1 0 1 1 6 8]
Using Deque: 21.873658ms
Using Brute Force: 6.042102028s
*/
```

The results are dramatic: **21.87ms** for the deque-based algorithm and **6.04 seconds** for the brute-force algorithm.

In the next section, we introduce and implement a priority queue.

6.7 Priority Queue

Priority queues exist in many real-world situations. For example, when passengers line up to board a plane, many airlines associate a priority with each passenger. This may be based on age (children enjoy high priority), price for the ticket (first-class passengers get high priority), loyalty points (frequent traveler), disability, or other factors that determine the customer's priority. Within each priority grouping, the usual FIFO queue rules apply.

We assume here that only a bounded number of priorities can be assigned to each item to be inserted in the queue.

We show one implementation in which we use a slice in which each element of the slice contains an ordinary queue.

The first queue in the slice contains items assigned the highest priority. The second queue in the slice contains items assigned the second highest priority and so on.

When an item is inserted, we access the queue corresponding to its priority and do an insertion in that queue.

Using a node-based queue for each element of the slice, we define a generic *PriorityQueue* and a function for creating the priority queue as follows:

```
type PriorityQueue[T any] struct {
    q []nodequeue.Queue[T] // slice of queues
    size int
}

func NewPriorityQueue[T any](numberPriorities int) (pq PriorityQueue[T]) {
    pq.q = make([]nodequeue.Queue[T], numberPriorities)
    return pq
}
```

The *NewPriorityQueue* constructor function defines a slice with *numberPriorities* node queues.

Listing 6-9 defines a *Passenger* type and presents an implementation of *PriorityQueue* along with a **main** driver. In the main driver, an airline queue with *Passenger* as the generic type is defined, and a group of passengers are inserted into the queue. Several passengers are removed, and the head of the line is output.

Listing 6-9. A slice implementation of priority queue and driver program

```
package main

import (
    "example.com/nodequeue"
    "fmt"
)

type Passenger struct {
    name string
    priority int
}
```

```go
type PriorityQueue[T any] struct {
    q []nodequeue.Queue[T]
    size int
}

func NewPriorityQueue[T any](numberPriorities int) (pq PriorityQueue[T]) {
    pq.q = make([]nodequeue.Queue[T], numberPriorities)
    return pq
}

// Methods for priority queue
func (pq *PriorityQueue[T]) Insert(item T, priority int) {
    pq.q[priority - 1].Insert(item)
    pq.size++
}

func (pq *PriorityQueue[T]) Remove() T {
    pq.size--
    for i := 0; i < len(pq.q); i++ {
        if pq.q[i].Size() > 0 {
            return pq.q[i].Remove()
        }
    }
    var zero T
    return zero
}

func (pq *PriorityQueue[T]) First() T {
    for _, queue := range(pq.q) {
        if queue.Size() > 0 {
            return queue.First()
        }
    }
    var zero T
    return zero
}
```

```go
func (pq *PriorityQueue[T]) IsEmpty() bool {
    result := true
    for _, queue := range(pq.q) {
        if queue.Size() > 0 {
            result = false
            break
        }
    }
    return result
}

func main() {
    airlineQueue := NewPriorityQueue[Passenger](3)
    passengers := []Passenger{ {"Erika", 3},{"Robert", 3}, {"Danielle", 3},
                               {"Madison", 1}, {"Frederik", 1}, {"James", 2},
                               {"Dante", 2}, {"Shelley", 3} }
    fmt.Println("Passsengers: ",passengers)
    for i := 0; i < len(passengers); i++ {
        airlineQueue.Insert(passengers[i], passengers[i].priority)
    }
    fmt.Println("First passenger in line: ", airlineQueue.First())
    airlineQueue.Remove()
    airlineQueue.Remove()
    airlineQueue.Remove()
    fmt.Println("First passenger in line: ", airlineQueue.First())
}
/* Output
Passsengers: [{Erika 3} {Robert 3} {Danielle 3} {Madison 1} {Frederik 1}
{James 2} {Dante 2} {Shelley 3}]
First passenger in line: {Madison 1}
First passenger in line after three Removes: {Dante 2}*/
```

The **Remove** method returns the zero value of **T** (**Passenger** in this case) if all the queues in the slice are empty.

The first three **Remove** invocations strip both priority 1 passengers from the queue and the first priority 2 passenger from the queue, making "Dante" the first in line.

We see in Listing 6-9 a layering of abstractions, a common practice in software development. We could have used a slice queue instead of the node queue by changing one line of code in the imports and changing each occurrence of ***nodequeue.Queue*** to ***slicequeue.Queue***.

In the next section, we present an important application of Queue – a discrete event simulation of a waiting line. A typical waiting line occurs when customers compete for service by lining up and waiting for a server to process each customer. An example would be the checkout process at a supermarket.

6.8 Queue Application: Discrete Event Simulation of Waiting Line

Suppose we have a waiting line for service at a bank. Customers arrive according to a **Poisson** arrival process with a specified average rate of arrival. Customers are served with a service time specified by a uniformly distributed random service time between a specified lower and upper bound. Our goal is to construct a simulation that estimates the average wait time (time from arrival on the line to time of completion of service) for a customer joining the line as well as other statistics taken over an eight-hour day.

Poisson Process

Events modeled by a Poisson arrival process satisfy the following conditions:

1.   Events are independent of each other. The occurrence of an event does not influence when another event occurs. If the events being modeled are customers arriving at a bank waiting line, this is probably a reasonable requirement to meet.

2.   The average rate of events remains constant. Here, we shall use a minute as the basic unit of time, so the average rate will be in events per minute.

It can be shown that the time between events, a random variable, can be generated using the function shown in the following:

```
func InterArrivalInterval(arrivalRate float64) float64 {
    // Models a Poisson process and returns
```

```
    rn := rand.Float64() // random float between 0.0 and 1
    return -math.Log(1.0 - rn) / arrivalRate
}
```

This corresponds to a probability that the wait time between events greater than some t is

P(Wait time > t) = e$^{-\lambda \ast t}$

where λ is the average arrival rate in events/minute. This is an exponential distribution. As t increases, the probability of the wait time exceeding **t** approaches 0. When **t** equals 0, the probability is 1. As the arrival rate λ increases (more events on average per minute), the probability of having the wait time for the next event to be greater than some **t** decreases.

We model the service duration (the time that it takes to process a customer) as a uniform distribution between **0.5 / arrival rate** and **1.4 / arrival rate**. So, for example, if the arrival rate is 0.25 (an average of one customer every 4 minutes), the service time is modeled as uniformly distributed between 2 minutes and 5.6 minutes or an average of 3.8 minutes. This leads to a stable queue since average service time is less than average time between arrivals.

Simulation Logic

Let us examine a typical sequence of events to set the stage for our simulation logic. The **a's** represent customer arrival times. The **d's** represent customer departure times. The line forms from left to right, so the leftmost customer is at the head of the line and is next to depart.

```
    t1
0   a1                              line: c1


        t2
0   a1  a2                          line: c1, c2


            t3
0   a1  a2      d1                  line: c2
    | <-  a1 service time -> |
```

				t4				
0	a1	a2		d1	a3			line: c2, c3

					t5			
0	a1	a2		d1	a3 d2			line: c3

						t6		
0	a1	a2		d1	a3 d2	d3		line: empty

The variable t (time) advances in discrete steps based on the next event – either an arrival or a departure – thus the name discrete-event simulation.

For the sequence of events shown, the wait times for customers 1, 2, and 3 are the following:

Customer 1: (d1 – a1)

Customer 2: (d2 – a2)

Customer 3: (d3 – a3)

The queue time is as follows: $(t2 - t1) * 1 + (t3 - t2) * 2 + (t4 - t3) * 1 + (t5 - t4) * 2 + (t6 - t5) * 1$

Average queue size: queue time / t6.

After each event (arrival or departure), the queue time is updated by taking the new event time – the last event time multiplied by the size of the queue. If the next event is a departure, the first customer on the queue is removed; its departure time – arrival time is added to the wait times. If the next event is an arrival, the customer is inserted into the queue.

The next arrival time is the previous arrival time + interval between arrivals. The next departure time is computed as the time when the customer becomes the first in the line + service time for the customer.

Implementation of System

With these observations in hand, we present Listing 6-10, which implements this system.

Listing 6-10. Discrete event simulation of waiting line

```
// Discrete event simulation of waiting line
package main

import (
```

```go
    "math/rand"
    "math"
    "fmt"
    "time"
    "example.com/nodequeue"
)

const (
    arrivalRate = 0.25 // average customer arrivals per minute
    lowerBoundServiceTime = 0.5 / arrivalRate
    upperBoundServicetime = 2.0 / arrivalRate
    quitTime = 480 // Minutes in an 8 hour day
)

func InterArrivalInterval(arrivalRate float64) float64 {
    // Models a Poisson process and returns
    rn := rand.Float64() // random float between 0.0 and 1
    return -math.Log(1.0 - rn) / arrivalRate
}

func ServiceTime() float64 {
    // Uniform distribution
    rn := rand.Float64() // rn between 0.0 and 1.0
    return lowerBoundServiceTime +
            (upperBoundServicetime - lowerBoundServiceTime) * rn
}

type Customer struct {
    arrivalTime float64
    serviceDuration float64
}

// ADT for Statistics
type Statistics struct {
    waitTimes []float64
    queueTime float64 // Accumulated time * queue size
```

```go
    longestQueue int
    longestWaitTime float64
}

func (s *Statistics) AddWaitTime(wait float64) {
    s.waitTimes = append(s.waitTimes, wait)
    if wait > s.longestWaitTime {
        s.longestWaitTime = wait
    }
}

func (s *Statistics) AddQueueSizeTime(queueSize int, timeAtSize float64) {
    s.queueTime += float64(queueSize) * timeAtSize
}

func (s *Statistics) AddLength(length int) {
    if length > s.longestQueue {
        s.longestQueue = length
    }
}

var lastArrivalTime, departureTime, lastEventTime float64

func main() {
    rand.Seed(time.Now().UnixNano())
    lastEventTime := 0.0 // beginning of day
    line := nodequeue.Queue[Customer]{}
    statistics := Statistics{}
    // Start simulation
    for {
        lastArrivalTime = lastArrivalTime + InterArrivalInterval(ar
        rivalRate)
        if lastArrivalTime > quitTime {
            break
        }
        if line.Size() == 0 {
            lastEventTime = lastArrivalTime
```

```
// fmt.Printf("\nline no longer empty at time: %0.2f. line size is 1",
lastEventTime)
            serviceTime := ServiceTime()
            customer := Customer{lastArrivalTime, serviceTime}
            line.Insert(customer)
            statistics.AddLength(line.Size())
            departureTime = lastArrivalTime + serviceTime
        } else {
            if lastArrivalTime < departureTime { // next event is an arrival
                customer := Customer{lastArrivalTime, ServiceTime()}
                statistics.AddQueueSizeTime(line.Size(), lastArrivalTime -
                            lastEventTime)
                lastEventTime = lastArrivalTime
                line.Insert(customer)
// fmt.Printf("\nArrival event at %0.2f - line size is: %d: ",
lastEventTime, line.Size())
                        statistics.AddLength(line.Size())
            } else { // next event is a departure
                statistics.AddQueueSizeTime(line.Size(), departureTime -
                            lastEventTime)
                departingCustomer := line.Remove()
                statistics.AddWaitTime(departureTime -
                            departingCustomer.arrivalTime)
                lastEventTime = departureTime
// fmt.Printf("\nDeparture event at %0.2f - line size is: %d: ",
lastEventTime, line.Size())
                if line.Size() > 0 {
                    departureTime = lastEventTime +
                                line.First().serviceDuration
                }
            }
        }
    }
```

```
    totalWaitTime := 0.0
    for i := 0; i < len(statistics.waitTimes); i++ {
        totalWaitTime += statistics.waitTimes[i]
    }
    averageWaitTime := totalWaitTime / float64(len(statistics.waitTimes))
    fmt.Printf("\nAverage Time from Arrival to Departure: %0.2f minutes",
                        averageWaitTime)
    fmt.Printf("\nAverage size of waiting line: %0.2f", statistics.
    queueTime / lastEventTime)
    fmt.Printf("\nLongest queue during the day: %d", statistics.
    longestQueue)
    fmt.Printf("\nLongest wait time during the day: %0.2f minutes",
                            statistics.longestWaitTime)
}
/* An output
Average Time from Arrival to Departure: 16.19 minutes
Average size of waiting line: 2.28
Longest queue during the day: 8
Longest wait time during the day: 40.18 minutes
*/
```

Multiple runs of the simulation exhibit a relatively large variance in the output statistics shown previously.

If we instrument the code with the three commented lines of code indented flush left, we output the exact sequence of events along with their event times.

A portion of the output produced during a typical run of the simulation is shown here:

```
line no longer empty at time: 0.81. line size is 1
Departure event at 6.23 - line size is: 0:
line no longer empty at time: 7.82. line size is 1
Departure event at 12.91 - line size is: 0:
line no longer empty at time: 28.79. line size is 1
Arrival event at 29.87 - line size is: 2:
Departure event at 33.56 - line size is: 1:
```

Departure event at 41.07 - line size is: 0:
line no longer empty at time: 53.59. line size is 1
Arrival event at 55.22 - line size is: 2:
Departure event at 58.17 - line size is: 1:
Departure event at 61.30 - line size is: 0:
line no longer empty at time: 62.79. line size is 1
Arrival event at 67.35 - line size is: 2:
Departure event at 70.29 - line size is: 1:
Arrival event at 71.87 - line size is: 2:
Departure event at 77.07 - line size is: 1:
Departure event at 81.35 - line size is: 0:
line no longer empty at time: 85.97. line size is 1
Arrival event at 89.44 - line size is: 2:
Departure event at 90.97 - line size is: 1:
Departure event at 93.27 - line size is: 0:
line no longer empty at time: 95.92. line size is 1
Departure event at 99.66 - line size is: 0:
line no longer empty at time: 105.60. line size is 1
Arrival event at 105.96 - line size is: 2:
Arrival event at 108.51 - line size is: 3:
Departure event at 109.23 - line size is: 2:
Arrival event at 114.78 - line size is: 3:
Arrival event at 115.19 - line size is: 4:
Arrival event at 115.93 - line size is: 5:
Departure event at 117.01 - line size is: 4:
Arrival event at 119.26 - line size is: 5:
Arrival event at 120.21 - line size is: 6:
Departure event at 124.32 - line size is: 5:
Departure event at 129.74 - line size is: 4:
Departure event at 133.90 - line size is: 3:
Departure event at 137.58 - line size is: 2:
Departure event at 140.13 - line size is: 1:
Departure event at 147.02 - line size is: 0:
line no longer empty at time: 170.46. line size is 1
Arrival event at 170.87 - line size is: 2:

```
Arrival event at 173.63 - line size is: 3:
Departure event at 177.92 - line size is: 2:
Arrival event at 181.92 - line size is: 3:
Departure event at 184.98 - line size is: 2:
Departure event at 192.90 - line size is: 1:
Departure event at 195.10 - line size is: 0:
line no longer empty at time: 211.92. line size is 1
Departure event at 215.48 - line size is: 0:
line no longer empty at time: 217.21. line size is 1
```

Many variations of the simulation presented here are interesting and useful. For example, we could investigate whether in the presence of multiple servers (e.g., bank tellers), it would be more efficient to have a single waiting line feeding all the tellers (line decreases in size by one when a teller is free) or separate waiting lines. We leave such investigations to the reader.

In the next section, we present another application of Queue: shuffling a deck of cards.

6.9 Queue Application: Shuffling Cards

In Section 3.4, Blackjack Game, we introduced the **Deck** abstraction as follows:

```go
var ranks = []string {"2", "3", "4", "5", "6", "7", "8", "9", "10", "J",
          "Q", "K", "A"}
var suits = []rune {'\u2660', '\u2661', '\u2662', '\u2663'}

type Card struct {
    Rank string
    Suit string
}

type Deck struct {
    Cards []Card
}
```

We saw that the **math/rand** package contains a **Shuffle()** method that can be applied to a slice, in this case, a slice of **Card**.

In this section, we construct our own **Shuffle** method using two queues.

Card Shuffling Model

Shuffling a deck of playing cards can be modeled as follows: Cut the deck of 52 cards into two piles, with a random size mismatch in the two piles of at most five cards. Then grab a card from alternating piles until the deck is reformed from the two separate piles. When the shorter pile has no more cards to contribute, add the cards from the larger pile directly to the deck.

If we model each pile as a queue of cards, then the shuffling process is straightforward.

Listing 6-11 presents the **Shuffle** method described earlier. We display the original deck and the shuffled deck.

Listing 6-11. Shuffling a deck of cards

```
// Shuffle deck of cards
package main

import (
    "example.com/nodequeue"
    "math/rand"
    "time"
    "fmt"
)

type Card struct {
    Rank string
    Suit string
}

type Deck struct {
    Cards []Card
}
```

```go
var ranks = []string {"2", "3", "4", "5", "6", "7", "8", "9", "10", "J",
        "Q", "K", "A"}
var suits = []rune {'\u2660', '\u2661', '\u2662', '\u2663'}

func NewDeck() (deck Deck) {
    for _, suit := range(suits) {
        for _, rank := range(ranks) {
            deck.Cards = append(deck.Cards, Card{rank, string(suit)})
        }
    }
    return deck
}

func (deck Deck) Shuffle() Deck {
    q1 := nodequeue.Queue[Card]{}
    q2 := nodequeue.Queue[Card]{}
    // Cut deck
    mismatch := -5 + rand.Intn(11) // -5 to 5
    var i int
    for i = 0; i < 26 + mismatch; i++ {
        q1.Insert(deck.Cards[i])
    }
    for ; i < 52; i++ {
        q2.Insert(deck.Cards[i])
    }
    // Rebuild deck
    deck = Deck{}
    for {
        if q1.Size() == 0 || q2.Size() == 0 {
            break
        }
        card := q1.Remove()
        deck.Cards = append(deck.Cards, card)
        card = q2.Remove()
        deck.Cards = append(deck.Cards, card)
    }
```

```go
        if q2.Size() == 0 {
            for {
                if q1.Size() == 0 {
                    break
                }
                card := q1.Remove()
                deck.Cards = append(deck.Cards, card)
            }
        }
        if q1.Size() == 0 {
            for {
                if q2.Size() == 0 {
                    break
                }
                card := q2.Remove()
                deck.Cards = append(deck.Cards, card)
            }
        }
        return deck
}

func main() {
    rand.Seed(time.Now().UnixNano())
    deck := NewDeck()
    fmt.Println("\nOriginal deck: ", deck)
    // Cut deck 5 times
    for index := 0; index < 5; index++ {
        deck = deck.Shuffle()
    }
    fmt.Println("\nShuffled deck: ", deck)
}
```

A typical output is shown as follows after five shuffles:

Original deck: {[{2 ♠} {3 ♠} {4 ♠} {5 ♠} {6 ♠} {7 ♠} {8 ♠} {9 ♠} {10 ♠} {J ♠} {Q ♠} {K ♠} {A ♠} {2 ♡} {3 ♡} {4 ♡} {5 ♡} {6 ♡} {7 ♡} {8 ♡} {9 ♡} {10 ♡} {J ♡} {Q ♡} {K ♡} {A ♡} {2 ◇} {3 ◇} {4 ◇} {5 ◇} {6 ◇} {7 ◇} {8 ◇} {9 ◇} {10 ◇} {J ◇} {Q ◇} {K ◇} {A ◇} {2 ♣} {3 ♣} {4 ♣} {5 ♣} {6 ♣} {7 ♣} {8 ♣} {9 ♣} {10 ♣} {J ♣} {Q ♣} {K ♣} {A ♣}]}

Shuffled deck: {[{2 ♠} {10 ♡} {Q ♡} {7 ♡} {9 ♣} {4 ♣} {6 ♣} {A ♠} {2 ♡} {J ◇} {K ◇} {8 ◇} {9 ♠} {4 ♠} {6 ♠} {A ♡} {3 ◇} {J ♣} {K ♣} {8 ♣} {9 ♡} {4 ♡} {6 ♡} {2 ♣} {3 ♣} {J ♠} {K ♠} {8 ♠} {10 ◇} {5 ◇} {7 ◇} {2 ◇} {3 ♠} {J ♡} {K ♡} {8 ♡} {10 ♣} {5 ♣} {7 ♣} {9 ◇} {3 ♡} {Q ◇} {A ◇} {5 ♠} {10 ♠} {Q ♣} {7 ♠} {5 ♡} {4 ◇} {Q ♠} {A ♣} {6 ◇}]}}

In method ***Shuffle***, the availability of a generic queue (in this case, from package **nodequeue** with generic parameter **Card**) greatly simplifies our work.

In the next section, we introduce the more general data structure, linked list.

6.10 Linked Lists

Lists play a fundamental role in software development. They hold a sequence of items from first to last. In a singly linked list, discussed in Section 6.11, each node containing an item points to the next node in the sequence. In a doubly linked list, discussed in Section 6.12, each node points forward to the next node in the list and backward to the previous node in the list. This allows us to traverse the list from first to last or from last to first.

Because of the linear structure of a list, it takes longer to access a particular item than in an array or slice. One needs to traverse the list, item by item, until the item being sought is found. In applications where fast direct access to an item through a location index is needed, arrays or slices are preferable.

We have already seen two specialized examples of linked lists: **nodestack and nodequeue**. In **nodestack**, information is inserted and removed from the leftmost node in the linked structure, assuming that elements are inserted from the left side with LIFO. In ***nodequeue***, information is inserted into the linked structure from left to right with FIFO.

In a linked list, information may be inserted anywhere (front, middle, end). Each insertion adds a new node to the list that is linked to the next node in a singly linked list or the next node and previous node in a doubly linked list. We shall show implementations of each.

The ADT operations that characterize a linked list are the following:

First( ) – Returns the first node in the list

Size( ) – Returns the number of nodes in the list

Insert(i, item) – Creates and inserts item in the ith node of the list

RemoveAt(i) – Removes and returns the item in the ith node of the list

Append(item) – Creates and inserts item into the last node of the list

IndexOf(item) – Returns the node position containing item in the list

Items( ) – Returns a slice of all the items in the list

In the next section, we present the implementation of a singly linked list.

6.11 Singly Linked List

A data structure for a generic singly linked list is given as follows:

```
package singlylinkedlist

import (
    "fmt"
)

type Ordered interface {
    ~string | ~int | ~float64
}

type Node[T Ordered] struct {
    Item T
    next *Node[T]
}

type List[T Ordered] struct {
    first       *Node[T]
    numberItems int
}
```

Type **Node** contains an **Item** (uppercase so it can be accessed outside the package) and a pointer to the next node in the list.

Type **List** contains a pointer, **first**, to the head node of the list and an integer **numberItems**.

Let us discuss in some detail two of the methods that can be invoked on List.

Method **Append** creates a new node and adds it to the end of the list as follows:

```
func (list *List[T]) Append(item T) {
    // Adds item to a new node at the end of the list
    newNode := Node[T]{item, nil}
    if list.first == nil {
        list.first = &newNode
    } else {
        last := list.first
        for {
            if last.next == nil {
                break
            }
            last = last.next
        }
        last.next = &newNode
    }
    list.numberItems += 1
}
```

A **newNode** with generic type **T** is defined with **item** and pointing to nil.

If the list the method is invoked on is empty, **list.first** is assigned to the address of **newNode**.

Otherwise, a for-loop is executed, advancing the pointer **last** until **last.next** is nil. Then **last.next** is assigned to the address of **newNode**.

Method **InsertAt** creates a new node and adds it at location **index** as follows:

```
func (list *List[T]) InsertAt(index int, item T) error {
    // Adds item to a new node at position index in the list
    if index < 0 || index > list.numberItems {
        return fmt.Errorf("Index out of bounds error")
    }
```

```
    newNode := Node[T]{item, nil}
    if index == 0 {
        newNode.next = list.first
        list.first = &newNode
        list.numberItems += 1
        return nil // No error
    }
    node := list.first
    count := 0
    previous := node
    for count < index {
        previous = node
        count++
        node = node.next
    }
    newNode.next = node
    previous.next = &newNode
    list.numberItems += 1
    return nil // no error
}
```

A test is first performed on the value of ***index*** and an error returned if ***index*** is less than 0 or greater than the number of existing items in the list.

As before, a new node is created with ***item*** and pointing to nil. If ***index*** is zero, new node is set to point to ***list.first***. Then ***list.first*** is assigned to the address of new node.

If index is not zero, a for-loop moves pointer ***node*** (initially assigned to ***list.first***) to the next node location with a trailing node, ***previous***, ***index – 1*** times.

Then two link assignments are made. The new node is inked to ***node***, and the ***previous*** node is assigned to the address of new node.

Listing 6-12 presents the entire package **singlylinkedlist**, and Listing 6-13 shows a driver program that exercises all the methods defined in Listing 6-12.

Listing 6-12. Package singlylinkedlist

```go
package singlylinkedlist

import (
    "fmt"
)

type Ordered interface {
    ~string | ~int | ~float64
}

type Node[T Ordered] struct {
    Item T
    next *Node[T]
}

type List[T Ordered] struct {
    first       *Node[T]
    numberItems int
}

// Methods
func (list *List[T]) Append(item T) {
    // Adds item to a new node at the end of the list
    newNode := Node[T]{item, nil}
    if list.first == nil {
        list.first = &newNode
    } else {
        last := list.first
        for {
            if last.next == nil {
                break
            }
            last = last.next
        }
```

```
        last.next = &newNode
    }
    list.numberItems += 1
}

func (list *List[T]) InsertAt(index int, item T) error {
    // Adds item to a new node at position index in the list
    if index < 0 || index > list.numberItems {
        return fmt.Errorf("Index out of bounds error")
    }
    newNode := Node[T]{item, nil}
    if index == 0 {
        newNode.next = list.first
        list.first = &newNode
        list.numberItems += 1
        return nil // No error
    }
    node := list.first
    count := 0
    previous := node
    for count < index {
        previous = node
        count++
        node = node.next
    }
    newNode.next = node
    previous.next = &newNode
    list.numberItems += 1
    return nil // no error
}

func (list *List[T]) RemoveAt(index int) (T, error) {
    if index < 0 || index > list.numberItems {
        var zero T
        return zero, fmt.Errorf("Index out of bounds error")
    }
```

```go
    node := list.first
    if index == 0 {
        toRemove := node
        list.first = toRemove.next
        list.numberItems -= 1
        return toRemove.Item, nil
    }
    count := 0
    previous := node
    for count < index {
        previous = node
        count++
        node = node.next
    }
    toRemove := node
    previous.next = toRemove.next
    list.numberItems -= 1
    return toRemove.Item, nil
}

func (list *List[T]) IndexOf(item T) int {
    node := list.first
    count := 0
    for {
        if node.Item == item {
            return count
        }
        if node.next == nil {
            return -1
        }
        node = node.next
        count += 1
    }
}
```

```go
func (list *List[T]) ItemAfter(item T) T {
    // Scan list for the first occurence of item
    node := list.first
    for {
        if node == nil { // item not found
            var zero T
            return zero
        }
        if node.Item == item {
            break
        }
        node = node.next
    }
    return node.next.Item
}

func (list *List[T]) Items() []T {
    result := []T{}
    node := list.first
    for i := 0; i < list.numberItems; i++ {
        result = append(result, node.Item)
        node = node.next
    }
    return result
}

func (list *List[T]) First() *Node[T] {
    return list.first
}

func (list *List[T]) Size() int {
    return list.numberItems
}
```

Listing 6-13. Main driver program for singlylinkedlist

```go
package main

import (
    "fmt"
    "example.com/singlylinkedlist"
)

func main() {
    cars := singlylinkedlist.List[string]{}
    cars.Append("Honda")
    cars.InsertAt(0, "Nissan")
    cars.InsertAt(0, "Chevy")
    cars.InsertAt(1, "Ford")
    cars.InsertAt(1, "Tesla")
    cars.InsertAt(0, "Audi")
    cars.InsertAt(2, "Volkswagon")
    cars.Append("Volvo")

    fmt.Println(cars.Items())
    fmt.Println("Index of Tesla: ", cars.IndexOf("Tesla"))

    cars.RemoveAt(0)
    car, _ := cars.RemoveAt(3)
    fmt.Println("car removed is: ", car)
    fmt.Println(cars.Items())
    cars.RemoveAt(cars.Size() - 1)
    fmt.Println(cars.Items())

    cars.Append("Lexus")
    fmt.Println(cars.Items())
    fmt.Println("First car in the list is: ", cars.First().Item)
    fmt.Println("Last car in the list is: ", cars.Items()[cars.Size() - 1])
}
/* Output
[Audi Chevy Volkswagon Tesla Ford Nissan Honda Volvo]
```

```
Index of Tesla:  3
car removed is:  Ford
[Chevy Volkswagon Tesla Nissan Honda Volvo]
[Chevy Volkswagon Tesla Nissan Honda]
[Chevy Volkswagon Tesla Nissan Honda Lexus]
First car in the list is:  Chevy
Last car in the list is:  Lexus
*/
```

It is left to the reader to examine and understand the remaining methods in package *singlylinkedlist* and verify that the output for the main driver program is correct.

In the next section, for completeness, we present the implementation details of a doubly linked list.

6.12 Doubly Linked List

In a doubly linked list, each node points to the previous as well as the next node in the list. This leads us to a data structure for a generic doubly linked list as follows:

```
type Ordered interface {
    ~string | ~int | ~float64
}

type Node[T Ordered] struct {
    Item T
    next *Node[T]
    prev *Node[T]
}

type List[T Ordered] struct {
    first        *Node[T]
    last         *Node[T]
    numberItems int
}
```

The **List** contains an additional field, *last*, that points to the end of the list. The methods for a doubly linked list are more complex because of the need to have each node link backward in addition to forward. These details are presented in Listing 6-14.

Listing 6-14. Package doublylinkedlist

```
package doublylinkedlist

import (
    "fmt"
)

type Ordered interface {
    ~string | ~int | ~float64
}

type Node[T Ordered] struct {
    Item T
    next *Node[T]
    prev *Node[T]
}

type List[T Ordered] struct {
    first       *Node[T]
    last        *Node[T]
    numberItems int
}

// Methods
func (list *List[T]) Append(item T) {
    // Adds item to a new node at the end of the list
    newNode := Node[T]{item, nil, nil}
    if list.first == nil {
        list.first = &newNode
        list.last = list.first
    } else {
        list.last.next = &newNode
        newNode.prev = list.last
        list.last = &newNode
```

```go
    }
    list.numberItems += 1
}

func (list *List[T]) InsertAt(index int, item T) error {
    // Adds item to a new node at position index in the list
    if index < 0 || index > list.numberItems {
        return fmt.Errorf("Index out of bounds error")
    }
    newNode := Node[T]{item, nil, nil}
    if index == 0 {
        newNode.next = list.first
        if list.first != nil {
            list.first.prev = &newNode
        }
        list.first = &newNode
        list.numberItems += 1
        if list.numberItems == 1 {
            list.last = list.first
        }
        return nil // No error
    }
    node := list.first
    count := 0
    previous := node
    for count < index {
        previous = node
        count++
        node = node.next
    }
    newNode.next = node
    previous.next = &newNode
    node.prev = &newNode
    newNode.prev = previous
```

```go
    list.numberItems += 1
    return nil // no error
}

func (list *List[T]) RemoveAt(index int) (T, error) {
    if index < 0 || index > list.numberItems {
        var zero T
        return zero, fmt.Errorf("Index out of bounds error")
    }
    node := list.first
    if index == 0 {
        toRemove := node
        list.first = toRemove.next
        list.numberItems -= 1
        if list.numberItems <= 1 {
            list.last = list.first
        }
        return toRemove.Item, nil
    }
    count := 0
    previous := node
    for count < index {
        previous = node
        count++
        node = node.next
    }
    toRemove := node
    previous.next = toRemove.next
    toRemove.next.prev = previous
    list.numberItems -= 1
    if list.numberItems <= 1 {
        list.last = list.first
    }
    return toRemove.Item, nil
}
```

```go
func (list *List[T]) IndexOf(item T) int {
    node := list.first
    count := 0
    for {
        if node.Item == item {
            return count
        }
        if node.next == nil {
            return -1
        }
        node = node.next
        count += 1
    }
}

func (list *List[T]) ItemAfter(item T) T {
    // Scan list for the first occurence of item
    node := list.first
    for {
        if node == nil { // item not found
            var zero T
            return zero
        }
        if node.Item == item {
            break
        }
        node = node.next
    }
    return node.next.Item
}

func (list *List[T]) ItemBefore(item T) T {
    // Scan list for the first occurence of item
    node := list.first
    for {
```

```go
        if node == nil { // item not found
            var zero T
            return zero
        }
        if node.Item == item {
            break
        }
        node = node.next
    }
    return node.prev.Item
}

func (list *List[T]) Items() []T {
    result := []T{}
    node := list.first
    for i := 0; i < list.numberItems; i++ {
        result = append(result, node.Item)
        node = node.next
    }
    return result
}

func (list *List[T]) ReverseItems() []T {
    result := []T{}
    node := list.last
    for {
        if node == nil {
            break
        }
        result = append(result, node.Item)
        node = node.prev
    }
    return result
}
```

```go
func (list *List[T]) First() *Node[T] {
    return list.first
}

func (list *List[T]) Last() *Node[T] {
    return list.last
}

func (list *List[T]) Size() int {
    return list.numberItems
}
```

If we compare the implementation details of this doubly linked list to the details presented earlier for the singly linked list, we see several assignment statements that link a node being added to the node that precedes it. When the list contains one node, the *first* and *last* pointers are the same.

We examine method ***InsertAt*** in detail. This is the most complex of all the methods. The other methods work in a similar way.

```go
func (list *List[T]) InsertAt(index int, item T) error {
    // Adds item to a new node at position index in the list
    if index < 0 || index > list.numberItems {
        return fmt.Errorf("Index out of bounds error")
    }
    newNode := Node[T]{item, nil, nil}
    if index == 0 {
        newNode.next = list.first
        if list.first != nil {
            list.first.prev = &newNode
        }
        list.first = &newNode
        list.numberItems += 1
        if list.numberItems == 1 {
            list.last = list.first
        }
        return nil // No error
    }
```

```
Node := list.first
count := 0
previous := node
for count < index {
    previous = node
    count++
    node = node.next
}
newNode.next = node
previous.next = &newNode
node.prev = &newNode
newNode.prev = previous
list.numberItems += 1
return nil // no error
}
```

After verifying that the index value sent in is not out of range, we create a **newNode** that contains item and points forward and backward to nil.

We first consider the case when index is 0. In this case, we link **newNode.next** to the existing **list.first**, even if it is nil (empty list). We then set **list.first** to **newNode** and increment list.numberItems. If the number of items is 1, we assign **list.last** to **list.first** since they are one in the same. We return nil to indicate no error.

In the case where **index** is not 0, we assign local variable **node** to **list.first**, local variable **count** to 0, and local variable **previous** to **node**.

In a for-loop that runs until **count** equals **index**, we set **previous** to **node**, increment **count**, and advance **node** to **node.next**.

Below the loop, having found the location in which to insert **newNode**, we do this insertion by setting **newNode.next** to **node**, setting **previous.next** to **&newNode**, setting **node.prev** to **&newNode**, setting **newNode.prev** to **previous**, and incrementing **list. numberItems**.

Benefit of Double Linking

What, you may be thinking, is the benefit of double linking considering the extra complexity required to construct and maintain a doubly linked list?

The most important benefit is the ability to traverse the list in reverse, from end to beginning. Although this is possible using a singly linked list, it would be computationally expensive.

6.13 Summary

In this chapter, we presented several important and useful generic data structures including **Queue**, **Deque**, **PriorityQueue**, and singly and doubly linked lists. We presented several applications that utilize these generic data structures.

In the next chapter, we present the hash table structure and some applications that use this structure.

CHAPTER 7

Hash Tables

In the previous chapter, we introduced queues and lists. We presented several specialized types of queues and their applications.

A map is a function that converts (maps) some **key** to a **value** in a **key-value** pair. The key and value may be of any type. A hash table is an unordered collection of key-value pairs where each key is distinct (no duplicate keys). Values are not required to be distinct, so two or more keys may map to the same value. Hash tables support very fast access to information accessed through keys. Fast table lookup is achieved by computing the hash value of a key and obtaining the location in the table containing the value.

In the next section, we review the standard Go data structure, map.

7.1 Map

A Go **map** type is declared as follows:

map[KeyType]ValueType

Listing 7-1 is a simple program that illustrates the basic operations using a standard **map**.

Listing 7-1. Using a standard map

```go
package main

import (
    "fmt"
)

type map1 map[string]string

func main() {
    nicknames := make(map1, 5)
    nicknames["Charles"] = "Chuck"
```

© Richard Wiener 2022
R. Wiener, *Generic Data Structures and Algorithms in Go*, https://doi.org/10.1007/978-1-4842-8191-8_7

```go
    nicknames["Robert"] = "Bob"
    nicknames["Richard"] = "Rick"
    nicknames["Teddy"] = "Ted"
    nicknames["Mohammad"] = "Mo"

    for key, value := range (nicknames) {
        fmt.Printf("\nThe nickname of %s is %s", key, value)
    }

    // Test for the presence of James in the map
    _, present := nicknames["James"]
    fmt.Println("\nThe key James is present: ", present)

    // Test for the presence of Teddy in the map
    _, present = nicknames["Teddy"]
    fmt.Println("The key Teddy is present: ", present)
    delete(nicknames, "Robert")

    // Test for the presence of Robert in the map
    _, present = nicknames["Robert"]
    fmt.Println("The key Robert is present: ", present)

    // Modify the nickname of Charles
    nicknames["Charles"] = "Charlie"
}
/* Output
The nickname of Robert is Bob
The nickname of Richard is Rick
The nickname of Teddy is Ted
The nickname of Mohammad is Mo
The nickname of Charles is Chuck
The key James is present:  false
The key Teddy is present:  true
The key Robert is present:  false
*/
```

The sequence of key-value output produced by the for-loop in Listing 7-1 may change from run to run of the program. A map is an unordered collection.

If a key is not present in a map, the value returned when testingx the key is the zero value associated with the value type stored in the map. But there are two return values when accessing a key. The first is the value associated with the key, and the second is a Boolean that determines the presence of the key in the map. We used this in several places in Listing 7-1 to determine whether a particular key is present in the map.

Hash Encryption

Hash function packages have been produced to support encryption. In Listing 7-2, we examine the use of two such hash-encryption packages.

Listing 7-2. Hash encryption

```
package main

import (
    "crypto/md5"
    "crypto/sha256"
    "fmt"
)

func main() {
    name1 := "Richard"
    name2 := "Richards"

    md5hash := md5.Sum([]byte(name1))
    sha256hash := sha256.Sum256([]byte(name1))
    fmt.Println("   MD5: ", md5hash)
    fmt.Println("SHA256: ", sha256hash)

    md5hash = md5.Sum([]byte(name2))
    sha256hash = sha256.Sum256([]byte(name2))
    fmt.Println("   MD5: ", md5hash)
    fmt.Println("SHA256: ", sha256hash)
}
```

```
/* Output
      MD5:[197 28 139 189 158 140 139 196 144 66 204 213 211 233 134 77]
SHA256:[29 235 10 59 134 117 13 14 74 76 33 220 150 1 115 105 84 174 92 202
198 84 197 127 61 69 86 58 31 89 89 152]
      MD5:[166 13 63 148 118 202 25 29 165 242 21 183 0 101 165 76]
SHA256:[107 180 140 197 199 134 66 52 247 101 104 172 63 77 46 205 135 103
147 106 45 109 84 183 195 48 107 144 11 99 127 198]
*/
```

The **Sum** method is invoked on **md5** with the input string converted to **[][byte**. The **Sum256** method is invoked on **sha256** with the input string converted to **[][byte**.

It is interesting to see how the addition of one character from **name1** to **name2** significantly changes the **md5** and **sha256** hash values.

These encryption functions are widely used to secure passwords.

In the next section, we examine the efficiency of a map and compare its search time with a slice.

7.2 How Fast Is a Map?

Suppose we examine the speed of accessing a large collection of dictionary words.

Specifically, we construct a **slice** of 466,551 English words taken from a text file. We next construct a **map** containing the same 466,551 words from the same text file. Then we compare the time it takes to look up every word in the slice vs. the map.

To speed up the **sliceCollection** search, we sort the words so we can use a binary search. We compare the map search time to the slice search time after we have sorted the words in the slice.

Listing 7-3 presents the details of this comparison.

Listing 7-3. Search time of map vs. slice

```
// We compare dictionary lookup using map versus slice
package main

import (
    "bufio"
    "fmt"
    "log"
```

```go
    "os"
    "sort"
    "time"
)

var mapCollection map[string]string

var sliceCollection []string

func IsPresent(word string, sliceCollection []string) bool {
    for i := 0; i < len(sliceCollection); i++ {
        if sliceCollection[i] == word {
            return true
        }
    }
    return false
}

func IsPresentBinarySearch(word string, sliceCollection []string) bool {
    // The slice collection is sorted
    low := 0
    high := len(sliceCollection) - 1
    for low <= high {
        median := (low + high) / 2

        if sliceCollection[median] < word {
            low = median + 1
        } else {
            high = median - 1
        }
    }
    if low == len(sliceCollection) || sliceCollection[low] != word {
        return false
    }
    return true
}
```

```go
func main() {
    file, err := os.Open("words.txt")
    defer file.Close()

    if err != nil {
        log.Fatalf("Error opening file: %s", err)
    }

    // Fill mapCollection and sliceConnection with words
    scanner := bufio.NewScanner(file)
    scanner.Split(bufio.ScanLines)

    mapCollection = make(map[string]string)
    sliceCollection = make([]string, 1)

    var words []string

    for scanner.Scan() {
        word := scanner.Text()
        words = append(words, word)
        mapCollection[word] = word
        sliceCollection = append(sliceCollection, word)
    }

    // Benchmark time to test for presence of each word in mapCollection
    start := time.Now()
    for i := 0; i < len(words); i++ {
        _, present := mapCollection[words[i]]
        if !present {
            fmt.Println("Word not found in mapCollectioOn")
        }
    }
    elapsed := time.Since(start)

    fmt.Println("Number of words in mapCollection: ", len(mapCollection))
    fmt.Println("\nTime to test words in mapCollection: ", elapsed)

    sort.Strings(sliceCollection)
```

```go
    // Benchmark time to test for presence of each word in sliceCollection
    start = time.Now()
    for i := 0; i < len(sliceCollection); i++ {
        if !IsPresent(sliceCollection[i], sliceCollection) {
            fmt.Println("Word not found in mapCollectioOn")
        }
    }
    elapsed = time.Since(start)
    fmt.Println("Time to test words in sliceCollection: ", elapsed)

    // Benchmark time to test for presence of each word in sorted
    // sliceCollection
    start = time.Now()
    for i := 0; i < len(sliceCollection); i++ {
        if !IsPresentBinarySearch(sliceCollection[i], sliceCollection) {
            fmt.Println("Word not found in mapCollectioOn")
        }
    }
    elapsed = time.Since(start)
    fmt.Println("Time to test words in sorted sliceCollection: ", elapsed)

}
/* Output
Number of words in mapCollection: 466468

Time to test words in mapCollection: 29.022542ms
Time to test words in sliceCollection: 2m20.874580833s
Time to test words in sorted sliceCollection: 51.836708ms
*/
```

The map-based collection is almost twice as fast as the sorted slice collection, confirming our earlier statement that maps provide very fast access to their information. The unsorted slice collection is inefficient and is many times slower than the map collection.

In the next section, we show how to build a hash table.

7.3 Building a Hash Table

Commercial-grade hash tables (maps) are complex and, as the previous section demonstrated, extremely efficient.

The details of how the Go map function is constructed are given in `https://github.com/golang/go/blob/master/src/runtime/map.go`.

In this section, we construct a relatively inefficient hash table from scratch so that we can briefly explore the issues related to hash-table construction.

Consider the following code segment:

```go
package main

import (
    "hash/fnv" // Fowler-Noll-Vo algorithm
    // Other details not shown yet
)

const tableSize = 100_000

func hash(s string) uint32 {
    h := fnv.New32a() // Fowler-Noll-Vo algorithm
    h.Write([]byte(s))
    return h.Sum32()
}

type WordType struct {
    word string
    list []string
}

// At every index there is a word and slice of words
type HashTable [tableSize]WordType
```

The function **hash** uses the Fowler-Noll-Vo algorithm to map a string to an unsigned 32-bit integer. The interested reader may wish to research the details of this algorithm.

We define a **WordType** as a struct with fields **word** and **list**.

Next, consider the function **NewTable** that creates and returns a **Hashtable**.

Create an Empty Hash Table

```
func NewTable() HashTable {
    var table HashTable
    for i := 0; i < tableSize; i++ {
        table[i] = WordType{"", []string{}}
    }
    return table
}
```

At every location in the **HashTable** array (fixed size **tableSize**), a variable of **WordType** is assigned with empty **word** and empty string slice, **[]string{}**.

Insertion into Hash Table

Now consider method **Insert** as follows:

```
func (table *HashTable) Insert(word string) {
    index := hash(word) % tableSize // Between 0 and tableSize - 1
    // Search table[index] for word
    if table[index].word == word {
        return // duplicates not allowed
    }
    if len(table[index].list) > 0 {
        for i := 0; i < len(table[index].list); i++ {
            if table[index].list[i] == word {
                return // duplicates not allowed
            }
        }
    }
    if table[index].word == "" {
        table[index].word = word
    } else {
        table[index].list = append(table[index].list, word)
    }
    length += 1
}
```

We "map" the input parameter, **word** (a string), to an index using the **hash** function displayed earlier.

A hash table cannot contain duplicate keys, so we test to see whether **word** already exists at **index**. If so, we return without changing the table.

We further test for a duplicate entry by scanning the **list**, if non-empty, for **word** and again return without changing the table if a duplicate is found.

Assuming the input word is not already in the table, we assign **word** to **table** at **index**. If there is already a word at the table of index, we append the input word to the string slice at location index.

Collisions and Collison Resolution

For a string slice at some index location to grow, there must be a collision. That is, the index assigned to the input word must collide with an existing word already at that location. Such collisions are inevitable since we compress **index** to be within the table size using

```
index := hash(word) % tableSize // Between 0 and tableSize - 1
```

Load Factor

The **load factor** of the table is the number of words in the table divided by the table size. Even if the load factor is less than 1, collisions may still occur because the hash function does not produce unique values for all input words. A table with many string slices (collision chains) at various index locations takes longer to access.

Determining Whether a Key Is Present

We next consider function **IsPresent** given as follows:

```
func (table HashTable) IsPresent(word string) bool {
    index := hash(word) % tableSize // Between 0 and tableSize - 1
    // Search table[index] for word
    if table[index].word == word {
        return true
    }
    if len(table[index].list) > 0 {
```

```
        for i := 0; i < len(table[index].list); i++ {
            if table[index].list[i] == word {
                return true
            }
        }
    }
    return false
}
```

This function returns true if the input word is at location *index* or in a non-empty list rooted at *index*.

Comparing the Performance of Hash Table with Standard Map

Listing 7-4 puts the pieces discussed earlier together and compares the execution time of the hash table with the standard map.

Listing 7-4. Comparing hash table to map

```
// Hash table construction
package main

import (
    "fmt"
    "hash/fnv" // Fowler-Noll-Vo algorithm
    "strconv"
    "time"
)

const tableSize = 100_000

var length int

func hash(s string) uint32 {
    h := fnv.New32a()
    h.Write([]byte(s))
    return h.Sum32()
}
```

```go
type WordType struct {
    word string
    list []string
}

// At every index there is a slice of words
type HashTable [tableSize]WordType

func NewTable() HashTable {
    var table HashTable
    for i := 0; i < tableSize; i++ {
        table[i] = WordType{"", []string{}}
    }
    return table
}

// Methods
func (table *HashTable) Insert(word string) {
    index := hash(word) % tableSize // Between 0 and tableSize - 1
    // Search table[index] for word
    if table[index].word == word {
        return // duplicates not allowed
    }
    if len(table[index].list) > 0 {
        for i := 0; i < len(table[index].list); i++ {
            if table[index].list[i] == word {
                return // duplicates not allowed
            }
        }
    }
    if table[index].word == "" {
        table[index].word = word
    } else {
        table[index].list = append(table[index].list, word)
    }
    length += 1
}
```

```go
func (table HashTable) IsPresent(word string) bool {
    index := hash(word) % tableSize // Between 0 and tableSize - 1
    // Search table[index] for word
    if table[index].word == word {
        return true
    }
    if len(table[index].list) > 0 {
        for i := 0; i < len(table[index].list); i++ {
            if table[index].list[i] == word {
                return true
            }
        }
    }
    return false
}

func main() {
    myTable := NewTable()
    mapCollection := make(map[string]string)

    words := []string{}
    for i := 0; i < 500_000; i++ {
        word := strconv.Itoa(i)
        words = append(words, word)
        myTable.Insert(word)
        mapCollection[word] = ""
    }

    fmt.Println("Benchmark test begins to test words: ", length)
    start := time.Now()
    for i := 0; i < length; i++ {
        if myTable.IsPresent(words[i]) == false {
            fmt.Println("Word not found in table: ", words[i])
        }
    }

    elapsed := time.Since(start)
```

```
    fmt.Println("Time to test all words in myTable: ", elapsed)

    start = time.Now()
    for i := 0; i < len(mapCollection); i++ {
        _, present := mapCollection[words[i]]
        if !present {
            fmt.Println("Word not found in mapCollection: ", words[i])
        }
    }
    elapsed = time.Since(start)
    fmt.Println("Time to test words in mapCollection: ", elapsed)
}

/* Output
Benchmark test begins to test words: 500000
Time to test all words in myTable: 1m17.880336666s
Time to test words in mapCollection: 24.405583ms
*/
```

A half-million words are generated by converting the integer index in a loop to a string and using the resulting strings as inputs to the table and to the map. The table takes almost 138 seconds to search for all the words that are entered compared to less than 25 milliseconds for the map. Quite a dramatic difference!

In the next section, we delve into the important application area of string searching. We present a classic string search algorithm that uses hashing as its basis.

7.4 Hash Application: String Search

We explore a classic and important string search application in this section. The **Rabin-Karp** algorithm features hashing and attempts to reduce the complexity of searching from **O(n*m)** to **O(n)**, where n is the length of the string to be searched and m is the length of the pattern we are searching for.

A function that accomplishes a string search using brute force is given as follows:

```
func BruteForceSearch(txt, pattern string) (bool, int) {
    patternLength := len(pattern)
    for outer := 0; outer < len(txt)-patternLength; outer++ {
        if txt[(outer):(outer+patternLength)] == pattern {
            return true, outer
        }
    }
    return false, -1
}
```

As an outer loop ranges from **0** to **len(txt) – patternLength**, we compare the string bounded by **txt[(outer):(outer+patternLength)]** to **pattern.** If the two strings are equal, we return true and the outer position. Since the string comparison is of **O(patternLength)** and we perform this operation n times, we have an **O(n * m)** algorithm, where m is the pattern length.

At this moment, you may rightfully be asking "what does this have to do with hashing?"

Suppose we replace the test for string equality, performed **n – m** times with a comparison of their hash values. That is, we compare **hash(txt[(outer):(outer+patternLength)])** with **hash(pattern)** and do this **n – m** times, returning true if their hash values are the same.

Rolling Hash Computation

What if the first hash computation, as **outer** index is incremented by one, can be determined from the previous hash value, avoiding the need to perform a separate hash from scratch? This is what the Rabin-Karp algorithm does. It uses a "rolling" hash function, where succeeding hash computations are inexpensively computed from the previous hash computation.

The hash function, **H**, is defined as follows for a portion of the text going from **i** to **i+ m – 1,** where m is the length of the pattern:

$$H_i = (c_i R^{m-1} + c_{i+1} R^{m-2} + ... + c_{i+m-1} R^0) \bmod Q$$

The c's are integer character values at the given locations in the string being searched, and R is a radix that corresponds to the number of possible values that each character can have. Q is a large prime number that serves to prevent the computed hash value from overflowing.

This function does not guarantee unique hash values for different strings, but if Q and n (the string length) are large, it minimizes the probability of a collision.

The hash value at location **i + 1** can be computed from the hash value at **i,** in constant time, as follows:

$$H_{i+1} = (H_i - c_iR^{m-1})R + c_{i+m}$$

Suppose we limit the character set to the numerals from "0" to "9". Our string search attempts to see whether a pattern defined by a string of numerals is contained in a larger string of numerals.

The hash function is given as follows:

```go
const (
    Radix = uint64(10)
    Q     = uint64(10 ^ 9 + 9)
)

func Hash(s string, Length int) uint64 {
    // Horner's method
    h := uint64(0)
    for i := 0; i < Length; i++ {
        h = (h*Radix + uint64(s[i])) % Q
    }
    return h
}
```

This is Horner's method for evaluating a polynomial. We use **uint64** for the integer values to avoid overflow.

Rabin-Karp Algorithm

The *Search* method uses the Rabin-Karp algorithm outlined earlier and is shown here:

```go
func Search(txt, pattern string) (bool, int) {
    strings.ToLower(txt)
    strings.ToLower(pattern)
    n := len(txt)
    m := len(pattern)
```

```
    patternHash := Hash(pattern, m)
    textHash := Hash(txt, m)
    if textHash == patternHash {
        return true, 0
    }
    PM := uint64(1)
    for i := 1; i <= m-1; i++ {
        PM = (Radix * PM) % Q
    }
    for i := m; i < n; i++ {
        textHash = (textHash + Q - PM*uint64(txt[i-m])%Q) % Q
        textHash = (textHash*Radix + uint64(txt[i])) % Q
        if (patternHash == textHash) && pattern == txt[(i-m+1):(i+1)] {
            return true, i - m + 1
        }
    }
    return false, -1
}
```

Since the equality of **patternHash** and **textHash** does not guarantee that the pattern has been found, we test the pattern against the segment of text to be sure.

Listing 7-5. Comparing Rabin-Karp to brute-force search

```
package main

import (
    "fmt"
    "strings"
    "time"
)

const (
    Radix = uint64(10)
    Q     = uint64(10 ^ 9 + 9)
)
```

```go
func BruteForceSearch(txt, pattern string) (bool, int) {
    patternLength := len(pattern)
    for outer := 0; outer < len(txt)-patternLength; outer++ {
        if txt[(outer):(outer+patternLength)] == pattern {
            return true, outer
        }
    }
    return false, -1
}

func Hash(s string, Length int) uint64 {
    // Horner's method
    h := uint64(0)
    for i := 0; i < Length; i++ {
        h = (h*Radix + uint64(s[i])) % Q
    }
    return h
}

func Search(txt, pattern string) (bool, int) {
    strings.ToLower(txt)
    strings.ToLower(pattern)
    n := len(txt)
    m := len(pattern)
    patternHash := Hash(pattern, m)
    textHash := Hash(txt, m)
    if textHash == patternHash {
        return true, 0
    }
    PM := uint64(1)
    for i := 1; i <= m-1; i++ {
        PM = (Radix * PM) % Q
    }
    for i := m; i < n; i++ {
        textHash = (textHash + Q - PM*uint64(txt[i-m])%Q) % Q
        textHash = (textHash*Radix + uint64(txt[i])) % Q
```

```
        if (patternHash == textHash) && pattern == txt[(i-m+1):(i+1)] {
            return true, i - m + 1
        }
    }
    return false, -1
}

func main() {
    text :="314159265358979323846264338327950288419716939937510582097494459230781640628620899862803482534211170679"
    pattern := "816406286208998628034825342"
    start := time.Now()
    _, _ = BruteForceSearch(text, pattern)
    elapsed := time.Since(start)
    fmt.Println("Computation time using BruteForceSearch: ", elapsed)

    start = time.Now()
    _, _ = Search(text, pattern)
    elapsed = time.Since(start)
    fmt.Println("Computation time using Search: ", elapsed)

    fmt.Println(BruteForceSearch(text, pattern))
    fmt.Println(Search(text, pattern))
}

/* Output with Macbook Pro using M1 Max
Computation time using BruteForceSearch: 10.083µs
Computation time using Search:  1.375µs
true 67
true 67

Using iMac with 3.2 GHz 8-Core Intel Xeon W
Computation time using BruteForceSearch: 354ns
Computation time using Search: 1.161µs
*/
```

The program was run on two computers, and the results are surprising. On the MacBook Pro with M1 Max processor and 32G of combined RAM, the Rabin-Karp search is over seven times faster in searching the first 100 digits of Pi, not surprising. On the iMac with a 3.2-GHz 8-core Xeon W processor, the opposite occurs. The brute-force algorithm returns a time that is over three times faster than the Rabin-Karp algorithm.

These contradictory benchmarks again highlight the fact that the hardware and instruction set of a particular machine can greatly influence the outcome of such a benchmark.

In the next section, we use hashing to implement a generic **Set**.

7.5 Generic Set

The Go language does not provide a **Set** data structure. In this section, we implement a generic **Set** data structure. A set stores unique values with ordering of the values.

The operations that define a Set are the following:

Insert(item) – Adds item to the existing set if not already present

Delete(item) – Removes item from the existing set if present

In(item) – Returns true if item is in the existing set, otherwise returns false

Items() – Returns a slice of items from the existing set

Size() – Returns the number of items in the existing set

Union(set2) – Returns all the unique items in the existing set and set2

Intersection(set2) – Returns all the items in both the existing set and set2

Difference(set2) – Returns the items in the existing set, not in set2

Subset(set2) – Returns true if all the items in set2 are in set1, otherwise false

---

In our package **set**, we define generic type **Set** as follows:

```
package set

type Ordered interface {
    ~string | ~int | ~float64
}

type Set[T Ordered] struct {
    items map[T]bool
}
```

Here, we define the *items* field of **Set** as a **map** with generic parameter **T** of type **Ordered**.

The **map** structure in Go requires that the key value type be ordered.

The *Insert* method is implemented as follows:

```
// Add item to set
func (set *Set[T]) Insert(item T) {
    if set.items == nil {
        set.items = make(map[T]bool)
    }
    // Prevent duplicate entry
    _, present := set.items[item]
    if !present {
        set.items[item] = true
    }
}
```

We first determine whether the *set* is empty, in which case *set.items* would be nil. In this case, we initialize the *map* using *make*.

To prevent a duplicate entry, which is not legal in a set, we test to see whether *item* is already in the *set*. If not, we assign item to the *set.items* map.

The *Delete* method is implemented as follows:

```
// Remove item from set
func (set *Set[T]) Delete(item T) {
    _, present := set.items[item]
```

```
    if present {
        delete(set.items, item)
    }
}
```

If the *item* is present, we delete it from the *set.items* map.

The *In* method is implemented as follows:

```
// Return true if item is in set, otherwise false
func (set *Set[T]) In(item T) bool {
    _, present := set.items[item]
    return present
}
```

We return true if item is in the *set.items* map, otherwise false.

The *Items* method is implemented as follows:

```
// Return a slice of all the items in set
func (set *Set[T]) Items() []T {
    items := []T{}
    for item := range set.items {
        items = append(items, item)
    }
    return items
}
```

We initialize an empty slice of type *T*. We iterate through the range of the *set.items* map and append each *item* to *items* which we return.

The complete package set is presented in Listing 7-6. The other methods are equally straightforward.

Listing 7-6. Package set

```
package set
type Ordered interface {
    ~string | ~int | ~float64
}
```

```go
type Set[T Ordered] struct {
    items map[T]bool
}

// Methods

// Add item to set
func (set *Set[T]) Insert(item T) {
    if set.items == nil {
        set.items = make(map[T]bool)
    }
    // Prevent duplicate entry
    _, present := set.items[item]
    if !present {
        set.items[item] = true
    }
}

// Remove item from set
func (set *Set[T]) Delete(item T) {
    _, present := set.items[item]
    if present {
        delete(set.items, item)
    }
}

// Return true if item is in set, otherwise false
func (set *Set[T]) In(item T) bool {
    _, present := set.items[item]
    return present
}

// Return a slice of all the items in set
func (set *Set[T]) Items() []T {
    items := []T{}
```

```go
    for item := range set.items {
        items = append(items, item)
    }
    return items
}

// Return the number of items in set
func (set *Set[T]) Size() int {
    return len(set.items)
}

// Return a new set containing all the unique items of set and set2
func (set *Set[T]) Union(set2 Set[T]) *Set[T] {
    result := Set[T]{}
    result.items = make(map[T]bool)
    for index := range set.items {
        result.items[index] = true
    }
    for j := range set2.items {
        _, present := result.items[j]
        if !present {
            result.items[j] = true
        }
    }
    return &result
}

// Return a new set containing the items found in both set and set2
func (set *Set[T]) Intersection(set2 Set[T]) *Set[T] {
    result := Set[T]{}
    result.items = make(map[T]bool)
    for i := range set2.items {
        _, present := set.items[i]
        if present {
            result.items[i] = true
        }
    }
}
```

```
        return &result
}

// Return a new set of items in set not found in set2
func (set *Set[T]) Difference(set2 Set[T]) *Set[T] {
    result := Set[T]{}
    result.items = make(map[T]bool)
    for i := range set.items {
        _, present := set2.items[i]
        if !present {
            result.items[i] = true
        }
    }
    return &result
}

// Return true if all items of set2 are in set
func (set *Set[T]) Subset(set2 Set[T]) bool {
    for i := range set.items {
        _, present := set2.items[i]
        if !present {
            return false
        }
    }
    return true
}
```

Listing 7-7 presents a main driver test program that exercises the methods of package set.

Listing 7-7. Main driver to exercise set package

```
package main

import (
    "example.com/set"
    "fmt"
)
```

```go
func main() {
    set1 := set.Set[int]{}
    set1.Insert(3)
    set1.Insert(5)
    set1.Insert(7)
    set1.Insert(9)
    set2 := set.Set[int]{}
    set2.Insert(3)
    set2.Insert(6)
    set2.Insert(8)
    set2.Insert(9)
    set2.Insert(11)
    set2.Delete(11)
    fmt.Println("Items in set2: ", set2.Items())

    fmt.Println("5 in set1: ", set1.In(5))
    fmt.Println("5 in set2: ", set2.In(5))

    fmt.Println("Union of set1 and set2: ", set1.Union(set2).Items())
    fmt.Println("Intersection of set1 and set2: ",
                    set1.Intersection(set2).Items())
    fmt.Println("Difference of set2 with respect to set1: ",
                    set2.Difference(set1).Items())
    fmt.Println("Size of this difference: ", set1.
Intersection(set2).Size())
}
/* Output
Items in set2: [6 8 9 3]
5 in set1:  true
5 in set2:  false
Union of set1 and set2: [9 3 5 7 6 8]
Intersection of set1 and set2: [3 9]
Difference of set2 with respect to set1: [6 8]
Size of this difference:  2
*/
```

7.6 Summary

We examined hash functions and hash tables in this chapter. We saw that hash tables using the standard map data structure are extremely efficient in searching an unordered collection. We looked at the classic Rabin-Karp for efficiently searching a string for a pattern. This algorithm uses a rolling hash function. Finally, we implemented a Set using a hash map.

In the next chapter, we turn our attention to Tree data structures. This is the first of several chapters that focus on binary trees.

CHAPTER 8

Binary Trees

In the previous chapter, we examined hash functions and hash tables and looked at several applications including string searching and the implementation of a Set that utilizes hashing.

In this chapter, we turn our attention to **Tree** structures. This is the first of several chapters that focus on trees. We introduce binary trees in this chapter. We look at mechanisms for traversing a binary tree. We tackle the challenging problem of graphically displaying a binary tree. To do this, we again use the third-party Fyne package to obtain the resources needed for such graphics.

In the next section, we define a binary tree.

8.1 Binary Trees

A binary tree is a specialized type of tree in which

- Every node has at most two children
- The children are called **left** and **right**

A binary tree with 7 nodes of height 5 with 3 leaf nodes is shown in Figure 8-1.

© Richard Wiener 2022
R. Wiener, *Generic Data Structures and Algorithms in Go*, https://doi.org/10.1007/978-1-4842-8191-8_8

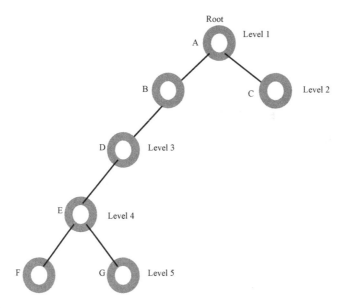

Figure 8-1. *Binary tree*

In contrast with trees in nature, the root of a binary tree is at the top of the structure, and the tree grows downward as more nodes are added. Leaf nodes have no children.

In the next section, we look at three methods for traversing a binary tree.

8.2 Tree Traversal

A tree traversal visits each node once operating on the data stored in each node. There are three traversals that we shall consider for binary trees: **inorder**, **preorder**, and **postorder**. Each of these traversals is defined recursively. We illustrate with the tree shown previously.

Inorder Traversal

Starting at the root, we descend left from A to B. We descend again left arriving at D. Then left again to E and finally again to F. No output has occurred yet. From F, we descend left only to find no left child. Then we backtrack and visit F. A visit could simply output the data stored in F or perform some operation on this data. We descend to the right of F again finding no right child. Having gone to F's left and outputting F and

F's right, we backtrack to E. Having already gone to E's left, we visit node E. We then descend to E's right. We go left from G; then we visit G. We backtrack to D. We visit D. We backtrack to B. We visit B. We backtrack to A. Having gone to A's left, we visit A. We descend to the right to node C. We go left. We visit C. We go right. We backtrack to A, and we are done.

The sequence of visitation is therefore **F, E, G, D, B, A, C**.

Preorder Traversal

For this traversal, we visit node A first. A visit could simply output the data stored in A or perform some operation on this data. After visiting A, we descend to the left reaching node B. We visit node B. Then we descend to the left and visit node D. Then we descend to the left and visit node E. Then we descend to the left and visit node F. Since node F does not have a left child, and having visited node F, we descend to the right and visit node G. We backtrack up the tree to node A. Having already visited A, we descend to the right and visit node C. The sequence of visitation is therefore **A B D E F G C**.

Postorder Traversal

Here, the recursive sequence of operations is descend left, descend right, and then visit. See whether you agree that this produces the sequence of visitation **F G E D B C A** for the tree shown previously.

In the next section, we implement a graphical depiction of a binary tree using the support of the third-party Fyne package.

8.3 Draw Tree

We wish to be able to draw a binary tree using graphics from the **fyne** graphical user interface package. Such a drawing must show all tree nodes with their key values and lines connecting parent and child nodes with the level of each node respected in the drawing.

Figure 8-2 shows a screenshot of the tree constructed from the code that we present in this section. The base type assumed for the data in each node is a single-character string.

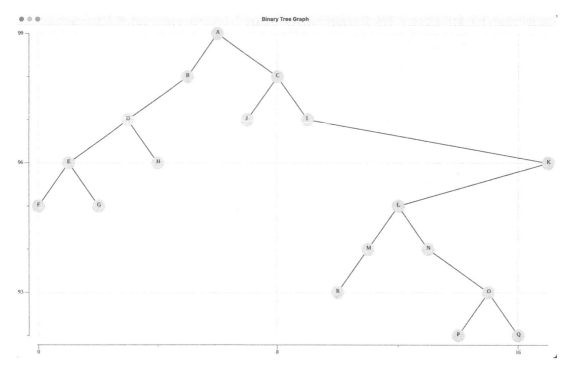

Figure 8-2. *Screenshot of a binary tree*

We simplify the explanation of the fairly complex draw-tree algorithm by first presenting a nongeneric version that uses **string** as the base type. In the next chapter, we present a generic version. Listing 8-1 presents the core data structures for drawing a binary tree.

Listing 8-1. Core data structures for drawing a binary tree

```
package main

type BinaryTree struct {
    Root *Node
    NumNodes int
}

type Node struct {
    Value string
    Left *Node
    Right *Node
```

```
}

type nodePair struct {
    Val1, Val2 string
}

type nodePos struct {
    Val string
    YPos int
    XPos int
}

var data []nodePos           // Used to get node positions (Val, XPos, YPos)
var endPoints []nodePair     // Used to plot lines
```

Binary Tree Structure

BinaryTree is defined as a struct containing a *Root* field specified as a pointer to a *Node* and a field *NumNodes*, an int.

Node is defined as a struct with a *Value* field of type string (later, it will be a generic type) and fields *Left* and *Right* each defined as a pointer to *Node*. A *Node* contains two recursive references to *Node* through *Left* and *Right* pointers.

Type *nodePair* is defined as a struct containing fields *Val1 and Val2* (string). Variable *endPoints* is defined as a slice of *nodePair* and is used to keep track of the end point values of the lines connecting nodes in the binary tree.

Type *nodePos* is a struct containing *Val* (string) and *YPos* and *XPos* (int). Variable *data* is defined as a slice of *nodePos* and is used to define the position and value of each node to be graphed.

Infrastructure Used to Display Binary Tree

Listing 8-2 shows the support functions that set up the infrastructure to display the graphics of the binary tree.

Listing 8-2. Functions for setting up the display of the binary tree

```
func prepareDrawTree(tree BinaryTree) {
    prepareToDraw(tree)
```

```go
        fmt.Println(endPoints)
        fmt.Println(data)
}

func findXY(val string) (int, int) {
    for i := 0; i < len(data); i++ {
        if data[i].Val == val {
            return data[i].XPos, data[i].YPos
        }
    }
    return -1, -1
}

func findX(val string) int {
    for i := 0; i < len(data); i++ {
        if data[i].Val == val {
            return i
        }
    }
    return -1
}

func setXValues() {
    for index := 0; index < len(data); index++ {
        xValue := findX(data[index].Val)
        data[index].XPos = xValue
    }
}

func prepareToDraw(tree BinaryTree) {
    inorderLevel(tree.Root, 1)
    setXValues()
    getEndPoints(tree.Root, nil)
}

func inorderLevel(node *Node, level int) {
    if node != nil {
        inorderLevel(node.Left, level + 1)
```

```
        data = append(data, nodePos{node.Value, 100 - level, -1})
        inorderLevel(node.Right, level + 1)
    }
}

func getEndPoints(node *Node, parent *Node) {
    if node != nil {
        if parent != nil {
            endPoints = append(endPoints, nodePair{node.Value,
            parent.Value})
        }
        getEndPoints(node.Left, node)
        getEndPoints(node.Right, node)
    }
}
```

Explanation of Code

The function *prepareDrawTree* invokes *prepareToDraw*, each taking a parameter *tree* (**BinaryTree**).

Function *prepareToDraw* invokes *inorderLevel* passing the root node of the tree and the level 1. This inorder traversal tests if the node is not nil and, if so, recursively calls itself passing the parameters *node.Left* and *level + 1*.

The second line of code is the visitation, which appends to the global *data* slice a *nodePos* with *node.Value* and *YPos* equal to 100 – level and an *XPos* of -1 (just a temporary place holder). Since trees are built from the root downward, the higher the level, the lower the *YPos*, thus the 100 – level for *YPos*.

The third line of code is the recursive call to *node.Right* and *level + 1*.

The second line of code in *prepareToDraw* is an invocation of the *setXValues()* function. This function uses *findX* to locate the index in the *data* slice that contains the value of every *nodePos* in *data*. This index is used as the *XValue* in the *nodePos* as we iterate through the *data* slice. The first *nodePos* in data will be the node furthest to the left (node F in the tree shown earlier) and will have an *XPos* of 0. The second *nodePos* in data will be node E in that tree. The slice *data* needs to be computed (except for *XPos*) before the *setXY()* function can do its work.

Upon the completion of *setXY()*, the preorder recursive function *getEndPoints* is invoked.

As each node is visited, the slice of *nodePair* is built using the node visited and its parent. This information will be used to draw the edges connecting the tree nodes.

We illustrate the construction of *data* and *endPoints* using a simple tree containing four nodes.

Listing 8-3 shows a simple **main** function and the resulting *data* and *endPoints* slices displayed in the console.

Listing 8-3. Main function with four nodes

```
package main

func main() {
    root := Node{"A", nil, nil}
    nodeB := Node{"B",nil, nil}
    nodeC := Node{"C", nil, nil}
    nodeD := Node{"D", nil, nil}

    root.Left = &nodeB
    root.Right = &nodeC
    nodeC.Right = &nodeD

    myTree := BinaryTree{&root, 4}
    ShowTreeGraph(myTree)
}
```

The console output is:

slice of endPoints: [{B A} {C A} {D C}]

slice of data: [{B 98 0} {A 99 1} {C 98 2} {D 97 3}]

The *data* slice reveals that node B has an *XPos* of 0 (leftmost node); node A, an *XPos* of 1; node C, an *XPos* of 2; and node D, an *XPos* of 3. This sequence results from the inorder traversal shown previously.

The end points of the three lines that must be drawn are shown in the slice of *endPoints* (a line from B to A, from C to A, and from D to C).

The tree that is constructed using the **fyne** GUI package is shown in Figure 8-3.

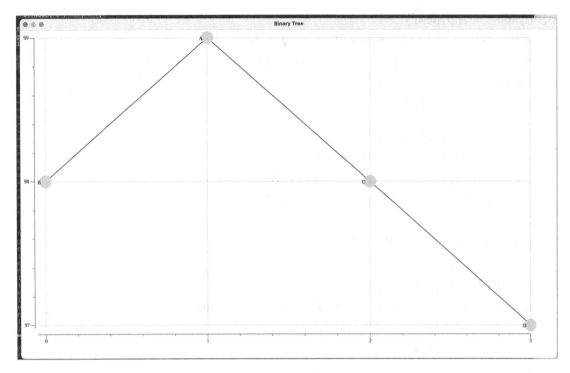

Figure 8-3. *Another binary tree screenshot*

Implementation of ShowTreeGraph

With the computation of the global variables ***data*** and ***endPoints***, we are ready to plot the graph representing the binary tree.

Listing 8-4 presents the details of plotting the tree graph.

Listing 8-4. Plotting the graph of the binary tree

```
func drawGraph(a fyne.App, w fyne.Window) {
    image := canvas.NewImageFromResource(theme.FyneLogo())
    image = canvas.NewImageFromFile(path + "tree.png")
    image.FillMode = canvas.ImageFillOriginal
    w.SetContent(image)
    w.Show()
}

func ShowTreeGraph(myTree BinaryTree) {
    prepareDrawTree(myTree)
```

273

```
myApp := app.New()
myWindow := myApp.NewWindow("Binary Tree")
myWindow.Resize(fyne.NewSize(1000, 600))
path, _ := homedir.Dir()
path += "/Desktop//"

nodePts := make(plotter.XYs, myTree.NumNodes)
for i := 0; i < len(data); i++ {
    nodePts[i].Y = float64(data[i].YPos)
    nodePts[i].X = float64(data[i].XPos)
}
nodePtsData := nodePts
p := plot.New()
p.Add(plotter.NewGrid())
nodePoints, err := plotter.NewScatter(nodePtsData)
if err != nil {
    log.Panic(err)
}
nodePoints.Shape = draw.CircleGlyph{}
nodePoints.Color = color.RGBA{G: 255, A: 255}
nodePoints.Radius = vg.Points(12)

// Plot lines
for index := 0; index < len(endPoints); index++ {
    val1 := endPoints[index].Val1
    x1, y1 := findXY(val1)
    val2 := endPoints[index].Val2
    x2, y2 := findXY(val2)
    pts := plotter.XYs{{X: float64(x1), Y: float64(y1)}, {X: float64(x2),
                            Y: float64(y2)}}
    line, err := plotter.NewLine(pts)
    if err != nil {
        log.Panic(err)
    }
    scatter, err := plotter.NewScatter(pts)
    if err != nil {
```

```go
            log.Panic(err)
        }
        p.Add(line, scatter)
    }

    p.Add(nodePoints)

    // Add Labels
    for index := 0; index < len(data); index++ {
        x := float64(data[index].XPos) - 0.05
        y := float64(data[index].YPos) - 0.02
        str := data[index].Val
        label, err := plotter.NewLabels(plotter.XYLabels {
            XYs: []plotter.XY {
                {X: x ,Y: y},
            },
            Labels: []string{str},
            },)
        if err != nil {
            log.Fatalf("Could not creates labels plotter: %+v", err)
        }
        p.Add(label)
    }

    path, _ = homedir.Dir()
    path += "/Desktop/GoDS/"
    err = p.Save(1000, 600, "tree.png")
    if err != nil {
        log.Panic(err)
    }

    drawGraph(myApp, myWindow)

    myWindow.ShowAndRun()
}
```

The first line of code in **ShowTreeGraph** is **prepareDrawTree**. This populates **data** with a slice of **nodePos**, with each **nodePos** containing the key value stored in a node as well as its **XPos** and **YPos** in the graph.

A new **fyne.Window, myWindow,** is created with the title "Binary Tree: and width 1000 and height 600 pixels".

A new plotter, **nodePts**, is created. The X and Y coordinates of the plotter are assigned from the **XPos** and **YPos** in the **data** slice.

A new **plot** is created and populated with the information in **plotter**. A scatter plot, **nodePoints**, is created from **plotter** using **nodePtsData**.

The **Shape**, **Color**, and **Radius** of each node point are assigned.

The same approach is taken in drawing the lines and creating the labels on each node.

Finally, a file "tree.png" is saved to the main directory.

The support function **drawGraph** is invoked with the fyne.App (**myApp**) and fyne. Window (**myWindow**) passed as parameters.

Function **drawGraph** loads and displays the "tree.png" image.

Many packages from the **fyne** framework need to be imported for the code to work. These imports are shown in Listing 8-5, which presents the complete **binarytree** package.

Listing 8-5. Complete binarytree package

```
package binarytree

import (
    "fmt"
    "image/color"
    "log"
    "fyne.io/fyne/v2"
    "fyne.io/fyne/v2/app"
    "fyne.io/fyne/v2/canvas"
    "fyne.io/fyne/v2/theme"
    "github.com/mitchellh/go-homedir"
    "gonum.org/v1/plot"
    "gonum.org/v1/plot/plotter"
    "gonum.org/v1/plot/vg"
    "gonum.org/v1/plot/vg/draw"
)
```

```go
type BinaryTree struct {
    Root *Node
    NumNodes int
}

type Node struct {
    Value string
    Left *Node
    Right *Node
}

type nodePair struct {
    Val1, Val2 string
}

type nodePos struct {
    Val string
    YPos int
    XPos int
}

var data []nodePos    // Used to get (Val, XPos, YPos) of each node
var endPoints []nodePair    // Used to plot lines

func prepareDrawTree(tree BinaryTree) {
    prepareToDraw(tree)
    fmt.Printf("\nslice of endPoints: %v", endPoints)
    fmt.Printf("\nslice of data: %v", data)
}

func findXY(val string) (int, int) {
    for i := 0; i < len(data); i++ {
        if data[i].Val == val {
            return data[i].XPos, data[i].YPos
        }
    }
    return -1, -1
}
```

```go
func findX(val string) int {
    for i := 0; i < len(data); i++ {
        if data[i].Val == val {
            return i
        }
    }
    return -1
}

func setXValues() {
    for index := 0; index < len(data); index++ {
        xValue := findX(data[index].Val)
        data[index].XPos = xValue
    }
}

func prepareToDraw(tree BinaryTree) {
    inorderLevel(tree.Root, 1)
    setXValues()
    getEndPoints(tree.Root, nil)
}

func inorderLevel(node *Node, level int) {
    if node != nil {
        inorderLevel(node.Left, level + 1)
        data = append(data, nodePos{node.Value, 100 - level, -1})
        inorderLevel(node.Right, level + 1)
    }
}

func getEndPoints(node *Node, parent *Node) {
    if node != nil {
        if parent != nil {
            endPoints = append(endPoints, nodePair{node.Value,
            parent.Value})
        }
```

```
            getEndPoints(node.Left, node)
            getEndPoints(node.Right, node)
    }
}

var path string

func drawGraph(a fyne.App, w fyne.Window) {
    image := canvas.NewImageFromResource(theme.FyneLogo())
    image = canvas.NewImageFromFile(path + "tree.png")
    image.FillMode = canvas.ImageFillOriginal
    w.SetContent(image)
    w.Show()
}

func ShowTreeGraph(myTree BinaryTree) {
    prepareDrawTree(myTree)
    myApp := app.New()
    myWindow := myApp.NewWindow("Binary Tree")
    myWindow.Resize(fyne.NewSize(1000, 600))
    path, _ := homedir.Dir()
    path += "/Desktop//"

    nodePts := make(plotter.XYs, myTree.NumNodes)
    for i := 0; i < len(data); i++ {
        nodePts[i].Y = float64(data[i].YPos)
        nodePts[i].X = float64(data[i].XPos)
    }
    nodePtsData := nodePts
    p := plot.New()
    p.Add(plotter.NewGrid())
    nodePoints, err := plotter.NewScatter(nodePtsData)
    if err != nil {
        log.Panic(err)
    }
    nodePoints.Shape = draw.CircleGlyph{}
    nodePoints.Color = color.RGBA{G: 255, A: 255}
    nodePoints.Radius = vg.Points(12)
```

```go
// Plot lines
for index := 0; index < len(endPoints); index++ {
    val1 := endPoints[index].Val1
    x1, y1 := findXY(val1)
    val2 := endPoints[index].Val2
    x2, y2 := findXY(val2)
    pts := plotter.XYs{{X: float64(x1), Y: float64(y1)},
                {X: float64(x2), Y: float64(y2)}}
    line, err := plotter.NewLine(pts)
    if err != nil {
        log.Panic(err)
    }
    scatter, err := plotter.NewScatter(pts)
    if err != nil {
        log.Panic(err)
    }
    p.Add(line, scatter)
}

p.Add(nodePoints)

// Add Labels
for index := 0; index < len(data); index++ {
    x := float64(data[index].XPos) - 0.05
    y := float64(data[index].YPos) - 0.02
    str := data[index].Val
    label, err := plotter.NewLabels(plotter.XYLabels {
        XYs: []plotter.XY {
            {X: x ,Y: y},
        },
        Labels: []string{str},
        },)
```

```
        if err != nil {
            log.Fatalf("Could not creates labels plotter: %+v", err)
        }
        p.Add(label)
    }

    path, _ = homedir.Dir()
    path += "/Desktop/GoDS/"
    err = p.Save(1000, 600, "tree.png")
    if err != nil {
        log.Panic(err)
    }

    drawGraph(myApp, myWindow)

    myWindow.ShowAndRun()
}
```

Listing 8-6 presents a **main** driver program that uses package **binarytree** to construct a **BinaryTree** with 18 nodes and then displays this tree.

Listing 8-6. A main driver program that builds and displays a binary tree

```
package main

import bt"example.com/binarytree"

func main() {
    root  := bt.Node{"A", nil, nil}
    nodeB := bt.Node{"B",nil, nil}
    nodeC := bt.Node{"C", nil, nil}
    nodeD := bt.Node{"D", nil, nil}
    nodeE := bt.Node{"E", nil, nil}
    nodeF := bt.Node{"F",nil, nil}
    nodeG := bt.Node{"G", nil, nil}
    nodeH := bt.Node{"H", nil, nil}
    nodeI := bt.Node{"I", nil, nil}
    nodeJ := bt.Node{"J", nil, nil}
    nodeK := bt.Node{"K", nil, nil}
    nodeL := bt.Node{"L", nil, nil}
```

```
    nodeM := bt.Node{"M", nil, nil}
    nodeN := bt.Node{"N", nil, nil}
    nodeO := bt.Node{"O", nil, nil}
    nodeP := bt.Node{"P", nil, nil}
    nodeQ := bt.Node{"Q", nil, nil}
    nodeR := bt.Node{"R", nil, nil}

    root.Left = &nodeB
    root.Right = &nodeC
    nodeB.Left = &nodeD
    nodeD.Right = &nodeH
    nodeD.Left = &nodeE
    nodeE.Left = &nodeF
    nodeE.Right = &nodeG
    nodeC.Right = &nodeI
    nodeC.Left = &nodeJ
    nodeI.Right = &nodeK
    nodeK.Left = &nodeL
    nodeL.Left = &nodeM
    nodeL.Right = &nodeN
    nodeN.Right = &nodeO
    nodeO.Left = &nodeP
    nodeO.Right = &nodeQ
    nodeM.Left = &nodeR
    myTree := bt.BinaryTree{&root, 18}
    bt.ShowTreeGraph(myTree)
}
```

The binary tree produced is shown in Figure 8-4.

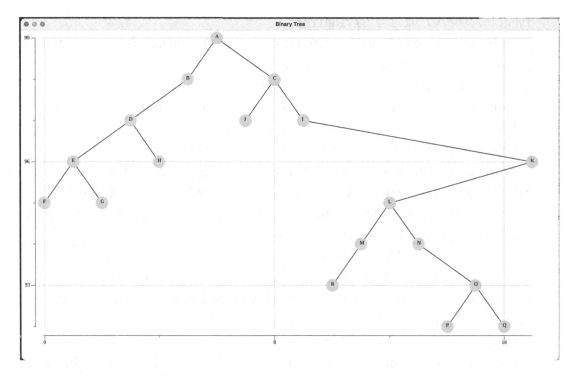

Figure 8-4. *Output of program*

Creating go.mod Files in Subdirectories binarytree and main

As discussed in Section 3.2, a module file, **go.mod**, must be generated in each of the subdirectories **main** and **binarytree** containing **main.go** and **binarytree.go**.

The invocation of **go mod tidy** in subdirectories **main** and **binarytree** causes the correct require clauses to be built in each of these **go.mod** files. The first time **main** is run, the imports from GitHub are downloaded.

The files generated are as follows:

```
module example.com/main

go 1.18

replace example.com/binarytree => ../binarytree

require (
    example.com/binarytree v0.0.0-00010101000000-000000000000 // indirect
```

```
    fyne.io/fyne/v2 v2.1.2 // indirect
    github.com/ajstarks/svgo v0.0.0-20210923152817-c3b6e2f0c527 // indirect
    github.com/davecgh/go-spew v1.1.1 // indirect
    github.com/fogleman/gg v1.3.0 // indirect
    github.com/fredbi/uri v0.0.0-20181227131451-3dcfdacbaaf3 // indirect
    github.com/fsnotify/fsnotify v1.4.9 // indirect
    github.com/go-fonts/liberation v0.2.0 // indirect
    github.com/go-gl/gl v0.0.0-20210813123233-e4099ee2221f // indirect
    github.com/go-gl/glfw/v3.3/glfw v0.0.0-20211024062804-40e447a793be
    github.com/go-latex/latex v0.0.0-20210823091927-c0d11ff05a81 // indirect
    github.com/go-pdf/fpdf v0.5.0 // indirect
    github.com/godbus/dbus/v5 v5.0.4 // indirect
    github.com/goki/freetype v0.0.0-20181231101311-fa8a33aabaff // indirect
    github.com/golang/freetype v0.0.0-20170609003504-e2365dfdc4a0 // indirect
    github.com/mitchellh/go-homedir v1.1.0 // indirect
    github.com/pmezard/go-difflib v1.0.0 // indirect
    github.com/srwiley/oksvg v0.0.0-20200311192757-870daf9aa564 // indirect
    github.com/srwiley/rasterx v0.0.0-20200120212402-85cb7272f5e9 // indirect
    github.com/stretchr/testify v1.5.1 // indirect
    github.com/yuin/goldmark v1.3.8 // indirect
    golang.org/x/image v0.0.0-20210628002857-a66eb6448b8d // indirect
    golang.org/x/net v0.0.0-20210405180319-a5a99cb37ef4 // indirect
    golang.org/x/sys v0.0.0-20210630005230-0f9fa26af87c // indirect
    golang.org/x/text v0.3.6 // indirect
    gonum.org/v1/plot v0.10.0 // indirect
    gopkg.in/yaml.v2 v2.2.8 // indirect
)

module example.com/binarytree

go 1.18

require (
    fyne.io/fyne/v2 v2.1.2
    github.com/mitchellh/go-homedir v1.1.0
    gonum.org/v1/plot v0.10.0
)
```

```
require (
    github.com/ajstarks/svgo v0.0.0-20210923152817-c3b6e2f0c527 // indirect
    github.com/davecgh/go-spew v1.1.1 // indirect
    github.com/fogleman/gg v1.3.0 // indirect
    github.com/fredbi/uri v0.0.0-20181227131451-3dcfdacbaaf3 // indirect
    github.com/fsnotify/fsnotify v1.4.9 // indirect
    github.com/go-fonts/liberation v0.2.0 // indirect
    github.com/go-gl/gl v0.0.0-20210813123233-e4099ee2221f // indirect
    github.com/go-gl/glfw/v3.3/glfw v0.0.0-20211024062804-40e447a793be
    github.com/go-latex/latex v0.0.0-20210823091927-c0d11ff05a81 // indirect
    github.com/go-pdf/fpdf v0.5.0 // indirect
    github.com/godbus/dbus/v5 v5.0.4 // indirect
    github.com/goki/freetype v0.0.0-20181231101311-fa8a33aabaff // indirect
    github.com/golang/freetype v0.0.0-20170609003504-e2365dfdc4a0 // indirect
    github.com/pmezard/go-difflib v1.0.0 // indirect
    github.com/srwiley/oksvg v0.0.0-20200311192757-870daf9aa564 // indirect
    github.com/srwiley/rasterx v0.0.0-20200120212402-85cb7272f5e9 // indirect
    github.com/stretchr/testify v1.5.1 // indirect
    github.com/yuin/goldmark v1.3.8 // indirect
    golang.org/x/image v0.0.0-20210628002857-a66eb6448b8d // indirect
    golang.org/x/net v0.0.0-20210405180319-a5a99cb37ef4 // indirect
    golang.org/x/sys v0.0.0-20210630005230-0f9fa26af87c // indirect
    golang.org/x/text v0.3.6 // indirect
    gopkg.in/yaml.v2 v2.2.8 // indirect
)
```

Now the import statement

```
package main

import bt"example.com/binarytree"
```

will work and allow the resources defined in package **binarytree** to be available in **main.go**.

8.4 Summary

We introduced the binary tree structure. We showed three mechanisms for visiting each tree node exactly once. We presented a nongeneric implementation of a binary tree and a suite of functions for graphically displaying a binary tree using the resources in the third-party **fyne** package.

In the next chapter, we continue our exploration of trees and examine binary search trees.

CHAPTER 9

Binary Search Tree

In the previous chapter, we introduced and implemented binary trees and explained the code for traversing and displaying such trees graphically.

In this chapter, we explore an important specialized binary tree, the binary search tree. The goal of a search tree is to organize data to support rapid access to the information stored in the tree. Search trees that are relatively balanced have a logarithmic relationship between the maximum depth of the tree and the number of nodes in the tree and therefore the number of operations required to search the tree for a particular item stored in the tree.

Tree search algorithms with complexity limited by maximum depth are highly efficient.

In the next section, we present an overview of search trees.

9.1 Overview

There are many types of search trees.

The first type of search tree we examine is the binary search tree. In later chapters, we explore other types of search trees.

A binary search tree (BST) is a special type of binary tree in which every node contains a search key and

1. All keys smaller than the key in node X are stored in the left subtree of X

2. All keys greater than the key in node X are stored in the right subtree of X

© Richard Wiener 2022

R. Wiener, *Generic Data Structures and Algorithms in Go*, https://doi.org/10.1007/978-1-4842-8191-8_9

This implies that in a search tree, we must be able to compare the **Value** fields of each node.

As an example, consider the BST shown in Figure 9-1. This tree is not balanced, but conditions 1 and 2 stated previously hold for every node in the tree.

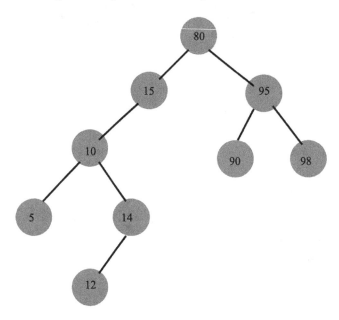

Figure 9-1. *Binary search tree*

Searching

Searching a binary search tree uses a simple algorithm. Compare the key you are searching for with the key in the root node. If the search key is smaller, descend to the left; if larger, descend to the right. Continue this pattern recursively until the bottom of the tree is reached or a node is found with a value that equals the search key.

As an example, if we were to search for node 12 in the tree shown previously, we would descend left (to node 15), descend left (to mode 10), descend right (to node 14), and descend left to our target, node 12. This would require five comparison operations, which is the depth of this tree.

Insertion

To insert a node into a search tree, we search for the node we wish to insert. This takes us to the bottom of the tree since we will not allow nodes with duplicate values in a search tree. We then insert the new node where it would have been found if initially present in the tree.

For the tree given earlier, if we were to insert node 13, it would be inserted as the right child of node 12 since that is where it would have been found if initially present in the tree.

Ordered Output

An inorder traversal of this search tree, and all search trees, produces a sequence of visitation from smallest value to largest value, ordered output. Try this out for the tree given in Figure 9-2 and verify this fact.

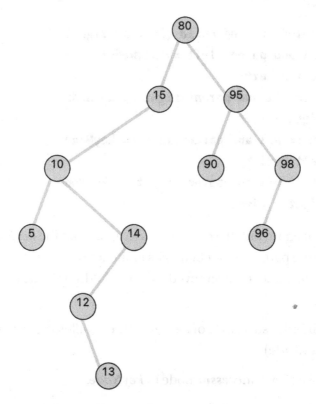

Figure 9-2. *Binary search tree for inorder traversal*

Deletion

The algorithm and method for removing a key from a search tree are more complicated. After the removal, we must be guaranteed to still have a search tree.

There are three special cases:

1.   The node to be removed is a leaf node.

2.   The node to be removed has one child.

3.   The node to be removed has two children.

Case 1 is the simplest. We find the parent of the leaf node and set the appropriate link (left or right) of the parent to nil, effectively clipping the leaf node from the tree.

The second case (the node to be removed has one child) is handled as a linked list deletion as follows:

Assume **left** is the left child of the node being deleted, or assume **right** is the right child of the node being deleted.

```
if node to be deleted has one child (left or right):
    if left != nil and parent.left == keyNode:
        parent.left = left
    else if left != nil and parent.right == keyNode:
        parent.right = left
    else if right != nil and parent.left == keyNode:
        parent.left = right
    else if right != nil and parent.right == keyNode:
        parent.right = right
```

As an exercise, diagram out these four cases to verify that the *keyNode* is unlinked from its parent and the parent reattached to its grandchild.

The third case (the node to be removed has two children) is the most complex. It is a three-step process:

1.   Find the inorder successor of *keyNode* (the smallest node to the right of *keyNode*).

2.   Copy the key from *successor* node to *keyNode*.

3.   Remove the *successor* node.

As an exercise, show that the *successor* node has either zero or one child, so its removal is a case 1 or case 2 deletion shown previously.

In the next section, we present a generic binary search tree implementation.

9.2 Generic Binary Search Tree

We present a generic implementation of binary search tree. We must constrain the generic type, **T**, so that it satisfies two conditions:

1. The values of type **T** stored in the tree nodes can be compared.

2. The values of type **T** stored in the tree nodes can be converted to a string using the **String( )** function.

Type OrderedStringer

We define a constraint type *OrderedStringer* that satisfies the two preceding conditions as follows:

```
type ordered interface {
    ~int | ~float64 | ~string
}

type OrderedStringer interface {
    ordered
    String() string
}
```

Requirement 1 is specified using the *ordered* type. Requirement 2 is specified using the signature for the *String()* function.

Any instantiation of the generic binary search tree must use a value type that satisfies the *OrderedStringer* type given previously. We will present examples later in this section that illustrates this usage.

Generic Types Needed for Binary Search Tree

Listing 9-1 presents the generic data structures needed in package **binarysearchtree**.

Listing 9-1. Generic data structures in package binarysearchtree

```go
package binarysearchtree
import (
    "image/color"
    "log"
    "fyne.io/fyne/v2"
    "fyne.io/fyne/v2/app"
    "fyne.io/fyne/v2/canvas"
    "fyne.io/fyne/v2/theme"
    "github.com/mitchellh/go-homedir"
    "gonum.org/v1/plot"
    "gonum.org/v1/plot/plotter"
    "gonum.org/v1/plot/vg"
    "gonum.org/v1/plot/vg/draw"
)

type ordered interface {
    ~int | ~float64 | ~string
}

type BinarySearchTree[T OrderedStringer] struct {
    Root *Node[T]
    NumNodes int
}

type Node[T OrderedStringer] struct {
    Value T
    Left *Node[T]
    Right *Node[T]
}

type OrderedStringer interface {
    ordered
    String() string
}
```

The type **BinarySearchTree** and the type **Node** are each defined with a generic parameter **T** of type **OrderedStringer**. This assures us that we can compare node values and output the values as strings in our graphical display of the generic tree. If we were not using the functions for displaying our binary search tree, we would not need the second constraint on the generic parameter **T**.

Methods for Binary Search Tree

The methods defined for type **BinarySearchTree** are presented in Listing 9-2.

Listing 9-2. Methods for BinarySearchTree

```
func (bst *BinarySearchTree[T]) Insert(newValue T) {
    if bst.Search(newValue) == false { // newValue not in existing tree
        n := &Node[T]{newValue, nil, nil}
        if bst.Root == nil { // First value in bst
            bst.Root = &Node[T]{newValue, nil, nil}
        } else {
            insertNode(bst.Root, n)
        }
        bst.NumNodes += 1
    }
}

func (bst *BinarySearchTree[T]) Delete(value T) {
    if bst.Search(value) == true {
        deleteNode(bst.Root, value)
        bst.NumNodes -= 1
    }
}

func (bst *BinarySearchTree[T]) Search(value T) bool {
    return search(bst.Root, value)
}

func (bst *BinarySearchTree[T]) InOrderTraverse(op func(T)) {
    inOrderTraverse(bst.Root, op)
}
```

```go
func (bst *BinarySearchTree[T]) Min() *T {
    node := bst.Root
    if node == nil {
        return nil
    }
    for {
        if node.Left == nil {
            return &node.Value
        }
        node = node.Left
    }
}

func (bst *BinarySearchTree[T]) Max() (*T, int) { // second return value is
                                                  // height
    node := bst.Root
    height := 1
    if node == nil {
        return nil, 0
    }
    for {
        if node.Right == nil {
            return &node.Value, height
        }
        height += 1
        node = node.Right
    }
}
```

Discussion of Insert, Delete, and Inorder Traversal

Methods **Insert** and **Delete** require that the node being inserted is not currently in the search tree and that the node being deleted is in the search tree. There is a small performance penalty imposed by this testing if the tree is relatively balanced.

The generic parameter constraint is not present in any of the methods. The compiler can infer this constraint since it is defined in type **BinarySearchTree**.

The method **InOrderTraversal** takes a function **op** as input. This represents the operation to be performed when visiting each node of the binary search tree.

Support Functions

Listing 9-3 contains the support functions that do the actual work defined in the publicly available methods defined in Listing 9-2.

Listing 9-3. Support functions for implementing methods

```
func insertNode[T OrderedStringer](node, newNode *Node[T]) {
    if newNode.Value < node.Value {
        if node.Left == nil {
            node.Left = newNode
        } else {
            insertNode(node.Left, newNode)
        }
    } else {
        if node.Right == nil {
            node.Right = newNode
        } else {
            insertNode(node.Right, newNode)
        }
    }
}

func deleteNode[T OrderedStringer](node *Node[T], value T) *Node[T] {
    if node == nil { return nil }
    if value < node.Value {
        node.Left = delete(node.Left, value)
        return node
    }
    if value> node.Value {
        node.Right = delete(node.Right, value)
        return node
    }
    if node.Left == nil && node.Right == nil {
        node = nil
        return nil
    }
```

```go
    if node.Left == nil {
        node = node.Right
        return node
    }
    if node.Right == nil {
        node = node.Left
        return node
    }
    LeftmostRightside := node.Right
    for {
        //find smallest value on the Right side
        if LeftmostRightside != nil && LeftmostRightside.Left != nil {
            LeftmostRightside = LeftmostRightside.Left
        } else {
            break
        }
    }
    node.Value = LeftmostRightside.Value
    node.Right = delete(node.Right, node.Value)
    return node
}

func search[T OrderedStringer](n *Node[T], value T) bool {
    if n == nil {
        return false
    }
    if value < n.Value {
        return search(n.Left, value)
    }
    if value > n.Value {
        return search(n.Right, value)
    }
    return true
}
```

```
func inOrderTraverse[T OrderedStringer](n *Node[T], op func(T)) {
    if n != nil {
        inOrderTraverse(n.Left, op)
        op(n.Value)
        inOrderTraverse(n.Right, op)
    }
}
```

Each of these support functions requires the explicit specification of the generic type constraint since the compiler cannot infer this from the function signature.

The support functions presented in Listing 9-3 are relatively simple recursive functions that perform the task indicated. It is left as an exercise for the reader to verify this.

Implementation of Tree Graphics

Listing 9-4 presents the code needed to display the binary search tree.

Listing 9-4. Code for graphing binary search tree

```
type NodePair struct {
    Val1, Val2 string
}

type NodePos struct {
    Val string
    YPos int
    XPos int
}

var data []NodePos
var endPoints []NodePair

func PrepareDrawTree[T OrderedStringer](tree BinarySearchTree[T]) {
    prepareToDraw(tree)
    // fmt.Println(endPoints)
    // fmt.Println(data)
}
```

```go
func FindXY(val interface{}) (int, int) {
    for i := 0; i < len(data); i++ {
        if data[i].Val == val {
            return data[i].XPos, data[i].YPos
        }
    }
    return -1, -1
}

func FindX(val interface{}) int {
    for i := 0; i < len(data); i++ {
        if data[i].Val == val {
            return i
        }
    }
    return -1
}

func SetXValues() {
    for index := 0; index < len(data); index++ {
        xValue := FindX(data[index].Val)
        data[index].XPos = xValue
    }
}

func prepareToDraw[T OrderedStringer](tree BinarySearchTree[T]) {
    inorderLevel(tree.Root, 1)
    SetXValues()
    getEndPoints(tree.Root, nil)
}

func inorderLevel[T OrderedStringer](node *Node[T], level int) {
    if node != nil {
        inorderLevel(node.Left, level + 1)
        data = append(data, NodePos{node.Value.String(), 100 - level, -1})
        inorderLevel(node.Right, level + 1)
    }
}
```

```go
func getEndPoints[T OrderedStringer](node *Node[T], parent *Node[T]) {
    if node != nil {
        if parent != nil {
            endPoints = append(endPoints, NodePair{node.Value.String(),
                    parent.Value.String()})
        }
        getEndPoints(node.Left, node)
        getEndPoints(node.Right, node)
    }
}

var path string

func DrawGraph(a fyne.App, w fyne.Window) {
    image := canvas.NewImageFromResource(theme.FyneLogo())
    image = canvas.NewImageFromFile(path + "tree.png")
    image.FillMode = canvas.ImageFillOriginal
    w.SetContent(image)
    w.Close()
    w.Show()
}

func ShowTreeGraph[T OrderedStringer](myTree BinarySearchTree[T]) {
    PrepareDrawTree(myTree)
    myApp := app.New()
    myWindow := myApp.NewWindow("Tree")
    myWindow.Resize(fyne.NewSize(1000, 600))
    path, _ := homedir.Dir()
    path += "/Desktop//"

    nodePts := make(plotter.XYs, myTree.NumNodes)
    for i := 0; i < len(data); i++ {
        nodePts[i].Y = float64(data[i].YPos)
        nodePts[i].X = float64(data[i].XPos)
    }
    nodePtsData := nodePts
    p := plot.New()
```

```go
p.Add(plotter.NewGrid())
nodePoints, err := plotter.NewScatter(nodePtsData)
if err != nil {
    log.Panic(err)
}
nodePoints.Shape = draw.CircleGlyph{}
nodePoints.Color = color.RGBA{G: 255, A: 255}
nodePoints.Radius = vg.Points(12)

// Plot lines
for index := 0; index < len(endPoints); index++ {
    val1 := endPoints[index].Val1
    x1, y1 := FindXY(val1)
    val2 := endPoints[index].Val2
    x2, y2 := FindXY(val2)
    pts := plotter.XYs{{X: float64(x1), Y: float64(y1)},
    {X: float64(x2), Y: float64(y2)}}
    line, err := plotter.NewLine(pts)
    if err != nil {
        log.Panic(err)
    }
    scatter, err := plotter.NewScatter(pts)
    if err != nil {
        log.Panic(err)
    }
    p.Add(line, scatter)
}

p.Add(nodePoints)

// Add Labels
for index := 0; index < len(data); index++ {
    x := float64(data[index].XPos) - 0.2
    y := float64(data[index].YPos) - 0.02
    str := data[index].Val
```

```
        label, err := plotter.NewLabels(plotter.XYLabels {
            XYs: []plotter.XY {
                {X: x ,Y: y},
            },
            Labels: []string{str},
            },)
        if err != nil {
            log.Fatalf("could not creates labels plotter: %+v", err)
        }
        p.Add(label)
    }

    path, _ = homedir.Dir()
    path += "/Desktop/GoDS/"
    err = p.Save(1000, 600, "tree.png")
    if err != nil {
        log.Panic(err)
    }

    DrawGraph(myApp, myWindow)
    myWindow.ShowAndRun()
}
```

The code follows the logic presented in Chapter 8 for displaying a binary tree.

Listings 9-5 and 9-6 present the complete code for package **binarysearchtree** and a main driver program that exercises the features of such a tree.

Listing 9-5. Package binarysearchtree

```
package binarysearchtree

import (
    "image/color"
    "log"
    "fyne.io/fyne/v2"
    "fyne.io/fyne/v2/app"
    "fyne.io/fyne/v2/canvas"
    "fyne.io/fyne/v2/theme"
    "github.com/mitchellh/go-homedir"
```

```go
    "gonum.org/v1/plot"
    "gonum.org/v1/plot/plotter"
    "gonum.org/v1/plot/vg"
    "gonum.org/v1/plot/vg/draw"
)

type ordered interface {
    ~int | ~float64 | ~string
}

type BinarySearchTree[T OrderedStringer] struct {
    Root *Node[T]
    NumNodes int
}

type Node[T OrderedStringer] struct {
    Value T
    Left *Node[T]
    Right *Node[T]
}

type OrderedStringer interface {
    ordered
    String() string
}

// Methods
func (bst *BinarySearchTree[T]) Insert(newValue T) {
    if bst.Search(newValue) == false { // newValue not in existing tree
        n := &Node[T]{newValue, nil, nil}
        if bst.Root == nil { // First value in bst
            bst.Root = &Node[T]{newValue, nil, nil}
        } else {
            insertNode(bst.Root, n)
        }
        bst.NumNodes += 1
    }
}
```

```go
func (bst *BinarySearchTree[T]) Delete(value T) {
    if bst.Search(value) == true {
        deleteNode(bst.Root, value)
        bst.NumNodes -= 1
    }
}

func (bst *BinarySearchTree[T]) Search(value T) bool {
    return search(bst.Root, value)
}

func (bst *BinarySearchTree[T]) InOrderTraverse(op func(T)) {
    inOrderTraverse(bst.Root, op)
}

func (bst *BinarySearchTree[T]) Min() *T {
    node := bst.Root
    if node == nil {
        return nil
    }
    for {
        if node.Left == nil {
            return &node.Value
        }
        node = node.Left
    }
}

func (bst *BinarySearchTree[T]) Max() (*T, int) { // second return value is
                                                 // height
    node := bst.Root
    height := 1
    if node == nil {
        return nil, 0
    }
```

```go
    for {
        if node.Right == nil {
            return &node.Value, height
        }
        height += 1
        node = node.Right
    }
}

// For internal use
func insertNode[T OrderedStringer](node, newNode *Node[T]) {
    if newNode.Value < node.Value {
        if node.Left == nil {
            node.Left = newNode
        } else {
            insertNode(node.Left, newNode)
        }
    } else {
        if node.Right == nil {
            node.Right = newNode
        } else {
            insertNode(node.Right, newNode)
        }
    }
}

func deleteNode[T OrderedStringer](node *Node[T], value T) *Node[T] {
    if node == nil {
        return nil
    }
    if value < node.Value {
        node.Left = deleteNode(node.Left, value)
        return node
    }
```

```go
    if value> node.Value {
        node.Right = deleteNode(node.Right, value)
        return node
    }
    if node.Left == nil && node.Right == nil {
        node = nil
        return nil
    }
    if node.Left == nil {
        node = node.Right
        return node
    }
    if node.Right == nil {
        node = node.Left
        return node
    }
    LeftmostRightside := node.Right
    for {
        //find smallest value on the Right side
        if LeftmostRightside != nil && LeftmostRightside.Left != nil {
            LeftmostRightside = LeftmostRightside.Left
        } else {
            break
        }
    }
    node.Value = LeftmostRightside.Value
    node.Right = deleteNode(node.Right, node.Value)
    return node
}

func search[T OrderedStringer](n *Node[T], value T) bool {
    if n == nil {
        return false
    }
```

```go
    if value < n.Value {
        return search(n.Left, value)
    }
    if value > n.Value {
        return search(n.Right, value)
    }
    return true
}

func inOrderTraverse[T OrderedStringer](n *Node[T], op func(T)) {
    if n != nil {
        inOrderTraverse(n.Left, op)
        op(n.Value)
        inOrderTraverse(n.Right, op)
    }
}

// Logic for drawing tree
type NodePair struct {
    Val1, Val2 string
}

type NodePos struct {
    Val string
    YPos int
    XPos int
}

var data []NodePos
var endPoints []NodePair // Used to plot lines

func PrepareDrawTree[T OrderedStringer](tree BinarySearchTree[T]) {
    prepareToDraw(tree)
    // fmt.Println(endPoints)
    // fmt.Println(data)
}
```

```go
func FindXY(val interface{}) (int, int) {
    for i := 0; i < len(data); i++ {
        if data[i].Val == val {
            return data[i].XPos, data[i].YPos
        }
    }
    return -1, -1
}

func FindX(val interface{}) int {
    for i := 0; i < len(data); i++ {
        if data[i].Val == val {
            return i
        }
    }
    return -1
}

func SetXValues() {
    for index := 0; index < len(data); index++ {
        xValue := FindX(data[index].Val)
        data[index].XPos = xValue
    }
}

func prepareToDraw[T OrderedStringer](tree BinarySearchTree[T]) {
    inorderLevel(tree.Root, 1)
    SetXValues()
    getEndPoints(tree.Root, nil)
}

func inorderLevel[T OrderedStringer](node *Node[T], level int) {
    if node != nil {
        inorderLevel(node.Left, level + 1)
        data = append(data, NodePos{node.Value.String(), 100 - level, -1})
        inorderLevel(node.Right, level + 1)
    }
}
```

```
func getEndPoints[T OrderedStringer](node *Node[T], parent *Node[T]) {
    if node != nil {
        if parent != nil {
            endPoints = append(endPoints, NodePair{node.Value.String(),
                            parent.Value.String()})
        }
        getEndPoints(node.Left, node)
        getEndPoints(node.Right, node)
    }
}

var path string

func DrawGraph(a fyne.App, w fyne.Window) {
    image := canvas.NewImageFromResource(theme.FyneLogo())
    image = canvas.NewImageFromFile(path + "tree.png")
    image.FillMode = canvas.ImageFillOriginal
    w.SetContent(image)
    w.Close()
    w.Show()
}

func ShowTreeGraph[T OrderedStringer](myTree BinarySearchTree[T]) {
    PrepareDrawTree(myTree)
    myApp := app.New()
    myWindow := myApp.NewWindow("Tree")
    myWindow.Resize(fyne.NewSize(1000, 600))
    path, _ := homedir.Dir()
    path += "/Desktop//"

    nodePts := make(plotter.XYs, myTree.NumNodes)
    for i := 0; i < len(data); i++ {
        nodePts[i].Y = float64(data[i].YPos)
        nodePts[i].X = float64(data[i].XPos)
    }
    nodePtsData := nodePts
    p := plot.New()
```

```go
p.Add(plotter.NewGrid())
nodePoints, err := plotter.NewScatter(nodePtsData)
if err != nil {
    log.Panic(err)
}
nodePoints.Shape = draw.CircleGlyph{}
nodePoints.Color = color.RGBA{G: 255, A: 255}
nodePoints.Radius = vg.Points(12)

// Plot lines
for index := 0; index < len(endPoints); index++ {
    val1 := endPoints[index].Val1
    x1, y1 := FindXY(val1)
    val2 := endPoints[index].Val2
    x2, y2 := FindXY(val2)
    pts := plotter.XYs{{X: float64(x1), Y: float64(y1)},
    {X: float64(x2),
                            Y: float64(y2)}}
    line, err := plotter.NewLine(pts) .
    if err != nil {
        log.Panic(err)
    }
    scatter, err := plotter.NewScatter(pts)
    if err != nil {
        log.Panic(err)
    }
    p.Add(line, scatter)
}

p.Add(nodePoints)

// Add Labels
for index := 0; index < len(data); index++ {
    x := float64(data[index].XPos) - 0.2 // Originall .05
    y := float64(data[index].YPos) - 0.02
    str := data[index].Val
    label, err := plotter.NewLabels(plotter.XYLabels {
```

```
            XYs: []plotter.XY {
                {X: x ,Y: y},
            },
            Labels: []string{str},
            },)
        if err != nil {
            log.Fatalf("could not create labels plotter: %+v", err)
        }
        p.Add(label)
    }

    path, _ = homedir.Dir()
    path += "/Desktop/GoDS/"
    err = p.Save(1000, 600, "tree.png")
    if err != nil {
        log.Panic(err)
    }

    DrawGraph(myApp, myWindow)

    myWindow.ShowAndRun()
}
```

Listing 9-6. Main driver program that uses binarysearchtree package

package main

```
import (
    bst"example.com/binarysearchtree"
    "math/rand"
    "time"
    "fmt"
)

// Satisfies OrderedStringer because of ~int
// Also satisfies OrderedStringer because of String() method below
type Number int
```

```go
func (num Number) String() string {
    return fmt.Sprintf("%d", num)
}

type Float float64

func (num Float) String() string {
    return fmt.Sprintf("%0.1f", num)
}

func inorderOperator(val Float) {
    fmt.Println(val.String())
}

func main() {
    rand.Seed(time.Now().UnixNano())
    // Generate a random search tree
    randomSearchTree := bst.BinarySearchTree[Float]{nil, 0}
    for i := 0; i < 30; i++ {
        rn := 1.0 + 99.0 * rand.Float64()
        randomSearchTree.Insert(Float(rn))
    }
    time.Sleep(3 * time.Second)
    bst.ShowTreeGraph(randomSearchTree)
    randomSearchTree.InOrderTraverse(inorderOperator)
    min := randomSearchTree.Min()
    max, _ := randomSearchTree.Max()
    fmt.Printf("\nMinimum value in random search tree is %0.1f  \nMaximum
                    value in random search tree is %0.1f", *min, *max)

    start := time.Now()
    tree := bst.BinarySearchTree[Number]{nil, 0}
    for val := 0; val < 100_000; val++ {
        tree.Insert(Number(val))
    }
    elapsed := time.Since(start)
    _, ht := tree.Max()
```

```
    fmt.Printf("\nTime to build BST tree with 100,000 nodes in sequential
            order: %s. Height of tree: %d", elapsed, ht)
}
/* Output
1.2
4.4
6.9
7.7
13.8
14.7
17.3
17.9
20.8
21.2
24.6
25.0
25.1
30.2
33.6
33.9
38.0
46.5
47.0
56.1
56.5
57.2
57.4
60.7
70.5
72.6
75.5
83.3
92.1
94.5
```

```
Minimum value in random search tree is 1.2
Maximum value in random search tree is 94.5
Time to build BST tree with 100,000 nodes in sequential order:
35.645312291s. Height of tree: 100000
*/
```

Discussion of binarysearchtree Package and Main Driver

The code for displaying a binary search tree in Listing 9-5 uses .**String()** in multiple
places since the type *T* is not known. This invocation of **String()** is boldfaced in that
listing.

There are two binary search trees used in the main driver. The generic types used are
Number and *Float*. Both of these types are implicitly of type *OrderedStringer* since they
have a **String()** function defined.

A binary search tree of base type **float64** is constructed with 30 nodes. Each
invocation produces a different tree. One such tree is shown in Figure 9-3.

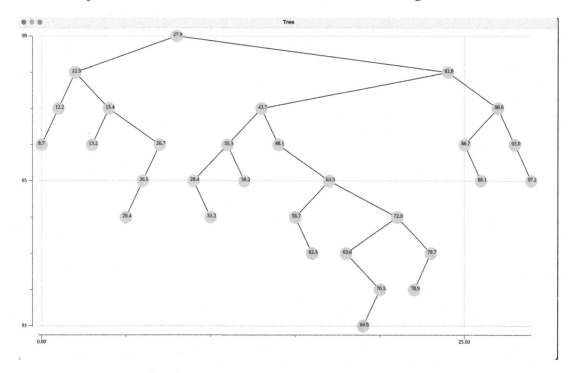

Figure 9-3. *A 30-node binary search tree*

This random binary search tree is unbalanced. The depth of the left subtree is 5, whereas the depth of the right subtree is 8.

The second binary search tree, with base type *Number*, is constructed by inserting 100,000 integers in sequential order. Essentially, this is a linked list. It took 35.6 seconds to build this completely unbalanced search tree.

9.3 Summary

In this chapter, we implemented a generic binary search tree. We made slight modifications to the draw tree logic so that the base type **T** could be output.

In the next chapter, we introduce one of the most important binary search trees, the AVL tree. This tree maintains its balance as new nodes are added to the tree.

## CHAPTER 10

AVL Trees

In the previous chapter, we introduced the binary search tree. In such a tree, each node contains a key that is larger than all the keys in its left subtree and smaller than all the keys in its right subtree. Duplicate keys are not allowed.

In building a binary search tree, the balance is dependent on the order in which keys are inserted. For example, if the keys are inserted in ascending order, the search tree resembles a linked list with its nodes to the right of the root node.

In 1962, two Russian mathematicians, Adelson Velsky and Landis, defined a useful definition of search tree balance (later called AVL balance in their honor) and described algorithms for *insert* and *remove* that preserve AVL balance. Their work has become a classic part of data structure legacy.

In the next section, we present an overview of AVL trees.

10.1 Overview: Adelson Velsky and Landis

In this chapter, we explore and implement a generic AVL tree.

For any binary search tree, the efficiency of *Insert*, *Delete*, and *Search* is dependent on how balanced the search is. Each of these operations requires approximately **$\log_2 n$** operations, if n is the number of nodes in the tree and the tree is balanced.

A binary search tree is defined as an **AVL** tree if **for every node in the tree, the maximum depth of the left subtree minus the maximum depth of the right subtree is equal or less than 1 in magnitude**. That is the AVL balance of every node is either -1, 0, or 1 and is given by the depth of the left subtree minus the depth of the right subtree.

In the following tree, the AVL balance (hereby called balance) of each node is shown. Balance 0 is not shown. This is not an AVL tree because of node A.

315

© Richard Wiener 2022
R. Wiener, *Generic Data Structures and Algorithms in Go*, https://doi.org/10.1007/978-1-4842-8191-8_10

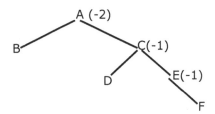

Please verify that the following search tree is an AVL tree.

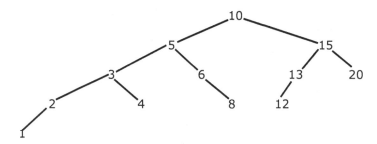

The AVL algorithms for insert and delete involve tree rotations. We illustrate with the previous tree.

Tree Rotations

A right rotate on node 10 yields

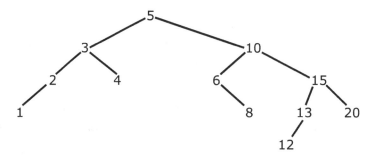

When 10 is rotated to the right, 5's right child becomes node 10. This makes 6 and 8 orphans. Since they are both greater than 5 and less than 10, they are placed into the left subtree of 10 and the right subtree of 5 as shown previously.

A left rotate on node 10 produces

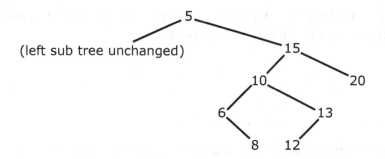

The orphan nodes 13 and 12 are larger than 10 and smaller than 15 and are placed as shown previously.

There is very little computational work involved in performing tree rotation. Only two links in the entire tree must be modified. This is true regardless of tree size.

The brilliance of the AVL algorithms is in the ***Insert*** and ***Delete*** methods. Both these methods are required to yield an AVL tree after either of these operations.

Insertion

We consider AVL insertion first. There are four steps.

1. Perform an ordinary binary search tree insertion into the AVL tree. If the tree is still an AVL tree, stop.

2. Starting at the node inserted (always at a leaf position), backtrack up the search path toward the root node. If a combination of nodes is found in which the parent has a balance whose absolute value is 2 and its child has a balance whose magnitude is 1, if the signs are the same (e.g., -2 and -1 or 2 and 1), a **type 1** configuration exists. If the signs are opposite (e.g., -2 and 1 or 2 and -1), a **type 2** configuration exists.

3. If the configuration is of type 1, perform a single rotation on the parent node in a direction to restore balance.

4. If the configuration is of type 2, perform a sequence of two rotations. The first rotation is on the child in a direction to restore balance. Then perform a second rotation in a direction opposite the first rotation on the parent.

These steps are guaranteed to produce a search tree with the AVL property intact. The proof of this is beyond the scope of this book.

Deletion

The steps for AVL deletion are given as follows:

1. Perform an ordinary binary search tree deletion. If the tree is an AVL tree, stop.

2. If the tree is not an AVL tree after the ordinary deletion, traverse up the search path from the node being deleted to the root node. Stop when one of the following combinations of balance occurs:

    a. Parent with balance whose absolute value is 2 and child with balance whose absolute value is 1. Determine the type of configuration as with insertion and perform the same type of either single rotation or sequence of two rotations.

    b. The parent has a balance of 2 or -2, and the child has a balance of 0. Consider this a type 1 configuration and perform the appropriate single rotation on the parent node.

    c. Reevaluate the balance of nodes above the parent. There is a possibility that because of the rotation(s) performed in step a or b, another configuration of type 1 or 2 needs to be dealt with. Continue step c until the root node is reached and no further rotational corrections are needed.

A tree that demonstrates the need for step c is shown in Figure 10-1.

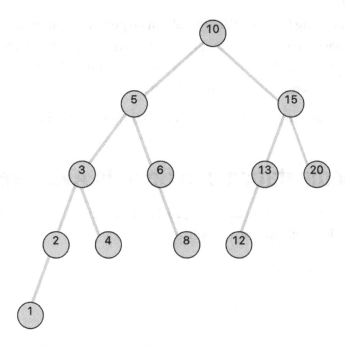

Figure 10-1. *A tree illustrating AVL deletion*

We wish to remove node 20. After the ordinary deletion of 20, node 15 has a balance of 2. This causes us to perform a right rotation on node 15. Node 13 moves upward and becomes the right child of the root node 10.

But now the root node 10 also has a balance of 2 since its left subtree is level 4 and its right subtree is level 2 (we lost a level in the right subtree during the first rotation). We correct this with a right rotation on node 10. As an exercise, sketch the resulting tree.

As an exercise, sketch the AVL tree resulting from step c.

Facts About AVL Trees

The following are some interesting facts about AVL trees:

- When inserting into AVL trees, approximately 50 percent of insertions require no rotational correction. Among the remaining 50 percent, about half require a type 1 single rotational correction, and half require the type 2 rotational corrections.

- When deleting from an AVL tree, about 80 percent require no
  rotational corrections. Among the remaining 20 percent, about
  half are type 1 and half type 2. Only rarely are multiple rotational
  corrections required up the search tree.

In the next section, we present an implementation of a generic AVL tree.

10.2 Implementation of a Generic AVL Tree

We present an entire **avl** package in Listing 10-1. Like the binary search tree, we include
the supporting code for displaying the AVL tree.

Listing 10-1. Package avl

```
package avl

import (
    "image/color"
    "log"
    "fyne.io/fyne/v2"
    "fyne.io/fyne/v2/app"
    "fyne.io/fyne/v2/canvas"
    "fyne.io/fyne/v2/theme"
    "github.com/mitchellh/go-homedir"
    "gonum.org/v1/plot"
    "gonum.org/v1/plot/plotter"
    "gonum.org/v1/plot/vg"
    "gonum.org/v1/plot/vg/draw"
)

type ordered interface {
    ~int | ~float64 | ~string
}

type AVLTree[T OrderedStringer] struct {
    Root *Node[T]
    NumNodes int
}
```

```go
type Node[T OrderedStringer] struct {
    Value T
    Left *Node[T]
    Right *Node[T]
    Ht int
}

type OrderedStringer interface {
    ordered
    String() string
}

// Methods
func (avl *AVLTree[T]) Insert(newValue T) {
    if avl.Search(newValue) == false { // newValue is not in existing tree
        avl.Root = insertNode(avl.Root, newValue)
        avl.NumNodes += 1
    }
}

func (avl *AVLTree[T]) Delete(value T) {
    if avl.Search(value) == true {
        avl.Root = deleteNode(avl.Root, value)
        avl.NumNodes -= 1
    }
}

func (avl *AVLTree[T]) Search(value T) bool {
    return search(avl.Root, value)
}

func (avl *AVLTree[T]) Height() int {
    return avl.Root.Height()
}

func (avl *AVLTree[T]) InOrderTraverse(f func(T)) {
    inOrderTraverse(avl.Root, f)
}
```

```
func (avl *AVLTree[T]) Min() *T {
    node := avl.Root
    if node == nil {
        return nil
    }
    for {
        if node.Left == nil {
            return &node.Value
        }
        node = node.Left
    }
}

func (avl *AVLTree[T]) Max() *T {
    node := avl.Root
    if node == nil {
        return nil
    }
    for {
        if node.Right == nil {
            return &node.Value
        }
        node = node.Right
    }
}

func (n *Node[T]) balanceFactor() int {
    if n == nil {
        return 0
    }
    return n.Left.Height() - n.Right.Height()
}

func (n *Node[T]) Height() int {
    if n == nil {
        return 0
    } else {
```

```
        return n.Ht
    }
}

func (n *Node[T]) updateHeight() {
    max := func (a, b int) int {
        if a > b {
            return a
        }
        return b
    }
    n.Ht = max(n.Left.Height(), n.Right.Height()) + 1
}

// Support functions
func newNode[T OrderedStringer](val T) *Node[T] {
    return &Node[T] {
        Value:  val,
        Left:   nil,
        Right:  nil,
        Ht: 1,
    }
}

func search[T OrderedStringer](n *Node[T], value T) bool {
    if n == nil {
        return false
    }
    if value < n.Value {
        return search(n.Left, value)
    }
    if value > n.Value {
        return search(n.Right, value)
    }
    return true
}
```

```go
func insertNode[T OrderedStringer](node *Node[T], val T) *Node[T] {
    // if there's no node, create one
    if node == nil {
        return newNode(val)
    }
    // if value is greater than current node's value, insert to the right
    if val > node.Value {
        right := insertNode(node.Right, val)
        node.Right = right
    }
    // if value is less than current node's value, insert to the left
    if val < node.Value {
        left:= insertNode(node.Left, val)
        node.Left = left
    }
    return rotateInsert(node, val)
}

func rightRotate[T OrderedStringer](x *Node[T]) *Node[T] {
    y := x.Left
    t := y.Right

    y.Right = x
    x.Left = t

    x.updateHeight()
    y.updateHeight()

    return y
}

func leftRotate[T OrderedStringer](x *Node[T]) *Node[T] {
    y := x.Right
    t := y.Left

    y.Left = x
    x.Right = t

    x.updateHeight()
```

```go
  y.updateHeight()

  return y
}

func rotateInsert[T OrderedStringer](node *Node[T], val T) *Node[T] {
  node.updateHeight()

  bFactor := node.balanceFactor()

  if bFactor > 1 && val < node.Left.Value {
    return rightRotate(node)
  }

  if bFactor < -1 && val > node.Right.Value {
    return leftRotate(node)
  }

  if bFactor > 1 && val > node.Left.Value {
    node.Left = leftRotate(node.Left)
    return rightRotate(node)
  }

  if bFactor < -1 && val < node.Right.Value {
    node.Right = rightRotate(node.Right)
    return leftRotate(node)
  }
  return node
}

func inOrderTraverse[T OrderedStringer](n *Node[T], op func(T)) {
    if n != nil {
        inOrderTraverse(n.Left, f)
        op(n.Value)
        inOrderTraverse(n.Right, f)
    }
}
```

```
func largest[T OrderedStringer](node *Node[T]) *Node[T] {
    if node == nil {
        return nil
    }

    if node.Right == nil {
        return node
    }
    return largest(node.Right)
}

func rotateDelete[T OrderedStringer](node *Node[T]) *Node[T] {
    node.updateHeight()
    bFactor := node.balanceFactor()

    if bFactor > 1 && node.Left.balanceFactor() >= 0 {
        return rightRotate(node)
    }

    if bFactor > 1 && node.Left.balanceFactor() < 0 {
        node.Left = leftRotate(node.Left)
        return rightRotate(node)
    }

    if bFactor < -1 && node.Right.balanceFactor() <= 0 {
        return leftRotate(node)
    }

    if bFactor < -1 && node.Right.balanceFactor() > 0 {
        node.Right = rightRotate(node.Right)
        return leftRotate(node)
    }
    return node
}

func deleteNode[T OrderedStringer](node *Node[T], val T) *Node[T] {
    if node == nil {
        return nil
    }
```

```
    if val > node.Value {
        right  := deleteNode(node.Right, val)
        node.Right = right
    } else if val < node.Value {
        left := deleteNode(node.Left, val)
            node.Left = left
    } else {
        if node.Left != nil && node.Right != nil {
            // has 2 children, find the successor
            successor := largest(node.Left)
            value := successor.Value

            // remove the successor
            left := deleteNode(node.Left, value)
            node.Left = left

            // copy the successor value to the current node
            node.Value = value
    } else if node.Left != nil || node.Right != nil {
            // has 1 child
            // move the child position to the current node
            if node.Left != nil {
            node = node.Left
            } else {
            node = node.Right
            }
    } else if node.Left == nil && node.Right == nil {
            // has no child
            // simply remove the node
            node = nil
    }
    if node == nil {
        return nil
    }
    return rotateDelete(node)
}
```

```go
// Logic for drawing tree
type NodePair struct {
    Val1, Val2 string
}

type NodePos struct {
    Val string
    YPos int
    XPos int
}

var data []NodePos
var endPoints []NodePair

func PrepareDrawTree[T OrderedStringer](tree AVLTree[T]) {
    prepareToDraw(tree)
    // fmt.Println(endPoints)
    // fmt.Println(data)

}


func FindXY(val interface{}) (int, int) {
    for i := 0; i < len(data); i++ {
        if data[i].Val == val {
            return data[i].XPos, data[i].YPos
        }
    }
    return -1, -1
}

func FindX(val interface{}) int {
    for i := 0; i < len(data); i++ {
        if data[i].Val == val {
            return i
        }
    }
    return -1
}
```

```go
func SetXValues() {
    for index := 0; index < len(data); index++ {
        xValue := FindX(data[index].Val)
        data[index].XPos = xValue
    }
}

func prepareToDraw[T OrderedStringer](tree AVLTree[T]) {
    inorderLevel(tree.Root, 1)
    SetXValues()
    getEndPoints(tree.Root, nil)
}

func inorderLevel[T OrderedStringer](node *Node[T], level int) {
    if node != nil {
        inorderLevel(node.Left, level + 1)
        data = append(data, NodePos{node.Value.String(), 100 - level, -1})
        inorderLevel(node.Right, level + 1)
    }
}

func getEndPoints[T OrderedStringer](node *Node[T], parent *Node[T]) {
    if node != nil {
        if parent != nil {
            endPoints = append(endPoints, NodePair{node.Value.String(),
                    parent.Value.String()})
        }
        getEndPoints(node.Left, node)
        getEndPoints(node.Right, node)
    }
}

var path string

func DrawGraph(a fyne.App, w fyne.Window) {
    image := canvas.NewImageFromResource(theme.FyneLogo())
    image = canvas.NewImageFromFile(path + "tree.png")
    image.FillMode = canvas.ImageFillOriginal
```

```
    w.SetContent(image)
    w.Close()
    w.Show()
}

func ShowTreeGraph[T OrderedStringer](myTree AVLTree[T]) {
    PrepareDrawTree(myTree)
    myApp := app.New()
    myWindow := myApp.NewWindow("Tree")
    myWindow.Resize(fyne.NewSize(1000, 600))
    path, _ := homedir.Dir()
    path += "/Desktop//"

    nodePts := make(plotter.XYs, myTree.NumNodes)
    for i := 0; i < len(data); i++ {
        nodePts[i].Y = float64(data[i].YPos)
        nodePts[i].X = float64(data[i].XPos)
    }
    nodePtsData := nodePts
    p := plot.New()
    p.Add(plotter.NewGrid())
    nodePoints, err := plotter.NewScatter(nodePtsData)
    if err != nil {
        log.Panic(err)
    }
    nodePoints.Shape = draw.CircleGlyph{}
    nodePoints.Color = color.RGBA{G: 255, A: 255}
    nodePoints.Radius = vg.Points(12)

    // Plot lines
    for index := 0; index < len(endPoints); index++ {
        val1 := endPoints[index].Val1
        x1, y1 := FindXY(val1)
        val2 := endPoints[index].Val2
        x2, y2 := FindXY(val2)
        pts := plotter.XYs{{X: float64(x1), Y: float64(y1)},
        {X: float64(x2), Y: float64(y2)}}
```

```
        line, err := plotter.NewLine(pts)
        if err != nil {
            log.Panic(err)
        }
        scatter, err := plotter.NewScatter(pts)
        if err != nil {
            log.Panic(err)
        }
        p.Add(line, scatter)
    }

    p.Add(nodePoints)

    // Add Labels
    for index := 0; index < len(data); index++ {
        x := float64(data[index].XPos) - 0.2 // Originall .05
        y := float64(data[index].YPos) - 0.02
        str := data[index].Val
        label, err := plotter.NewLabels(plotter.XYLabels {
            XYs: []plotter.XY {
                {X: x ,Y: y},
            },
            Labels: []string{str},
            },)
        if err != nil {
            log.Fatalf("could not creates labels plotter: %+v", err)
        }
        p.Add(label)
    }

    path, _ = homedir.Dir()
    path += "/Desktop/GoDS/"
    err = p.Save(1000, 600, "tree.png")
    if err != nil {
        log.Panic(err)
    }
    DrawGraph(myApp, myWindow)
    myWindow.ShowAndRun()
}
```

Explanation of avl Package

There are many functions to dissect. The easiest way to do this is using a debugger. I have used **VS Code** and **IntelliJ IDEA Ultimate**, which has a Go plug-in and an outstanding debugger.

We construct the tree shown earlier using the main driver code given in Listing 10-2.

Listing 10-2. Main driver code

```go
package main

import (
    avl "example.com/avl"
    "fmt"
)

type Integer int

func (num Integer) String() string {
    return fmt.Sprintf("%d", num)
}

func main() {
    myTree := avl.AVLTree[Integer]{nil, 0}
    myTree.Insert(10)
    myTree.Insert(15)
    myTree.Insert(5)
    myTree.Insert(3)
    myTree.Insert(6)
    myTree.Insert(13)
    myTree.Insert(20)
    myTree.Insert(2)
    myTree.Insert(4)
    myTree.Insert(8)
    myTree.Insert(12)
    myTree.Insert(1)
```

```
    // myTree.Delete(20)
    avl.ShowTreeGraph(myTree)
}
```

This produces the tree display shown in Figure 10-2.

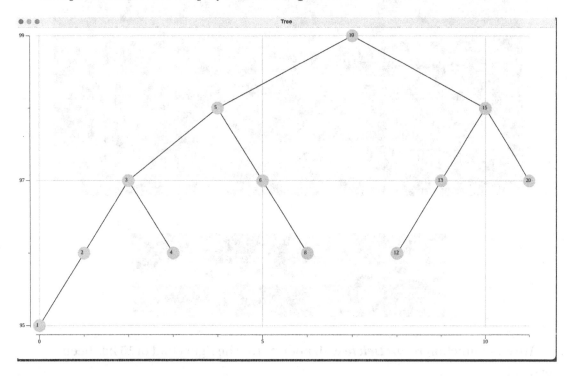

Figure 10-2. *Resulting AVL tree*

Using a debugger, let us "walk" through the code for deleting node 20, one of the harder use cases. If you do not have a debugger, just perform the "walk" visually, line by line.

We uncomment the line of code, **myTree.Delete(20)**, and set a break point at this line of code using **IntelliJ IDEA**.

```
44          myTree.Insert( newValue: 1)
45    ✓     myTree.Delete( value: 20)
46          avl.ShowTreeGraph(myTree)
```

We recursively descend to the right down the tree in function **deleteNode** until node is equal to nil. Then the recursion backtracks to node equal to 15.

```
v   P node = {*avl.Node[main.Integer] | 0xc000080c60}
        f Value = {main.Integer} 15
    v   f Left = {*avl.Node[main.Integer] | 0xc000080ce0}
            f Value = {main.Integer} 13
        >   f Left = {*avl.Node[main.Integer] | 0xc000080d80}
            f Right = {*avl.Node[main.Integer] | 0x0} nil
            f Ht = {int} 2
        f Right = {*avl.Node[main.Integer] | 0x0} nil
        f Ht = {int} 3
```

```
            }
        return rotateDelete(node)
    }
```

We enter function **rotateDelete** with node at 15. The right child of 15 has been set to nil.

The **bFactor** (balance) of node 15 is 2, and its left node has a **bFactor** of 1.

```
if bFactor > 1 && node.Left.balanceFactor() >= 0 {
    return rightRotate(node)
}
```

We invoke **rightRotate(node)**, where node is 15. The variable **y** gets set to 13; the right child of 13 gets set to 15. Node 13 is returned up the recursive chain (as **right** in the debugger code shown in the following).

```
if val > node.Value {
    right  := deleteNode(node.Right, val)
    node.Right = right
} else if val < node.Value {
```

Node 10 assigns its right child to 13. The return statement at the end of *deleteNode* returns the result of *rotateDelete(10)*.

```
if bFactor > 1 && node.Left.balanceFactor() >= 0 {
    return rightRotate(node)
}
```

Based on the *bFactor* of 10 (greater than 1) and the *bFactor* of 5 (equal or greater than 0), we next perform a *rightRotate(10)*.

The value of 5 is returned up the chain and becomes the new root node of the tree. The right child of 5 becomes 10. The left child of 10 becomes 6.

The new tree is shown in Figure 10-3.

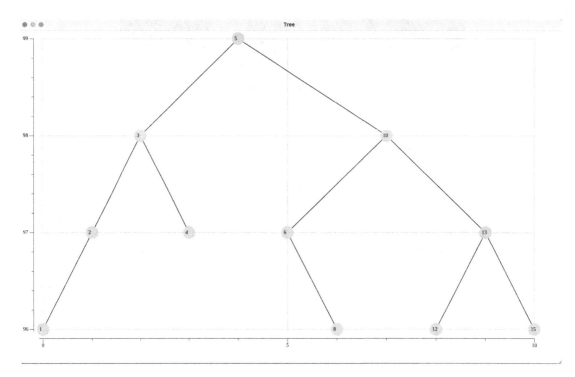

Figure 10-3. *Tree resulting from deletion*

Some additional tests of AVL trees are presented in Listing 10-3 (uncomment the test you wish to perform).

Listing 10-3. Another main driver with more AVL tests

```
package main

import (
    avl "example.com/avl"
    "fmt"
    "math/rand"
    "time"
)

func inorderOperator(val Float) {
    val *= val
    fmt.Println(val.String())
}
```

```go
// Satisfies OrderedStringer because of ~float64
// Also satisfies OrderedStringer because of String() method below
type Float float64

func (num Float) String() string {
    return fmt.Sprintf("%0.1f", num)
}

type Integer int

func (num Integer) String() string {
    return fmt.Sprintf("%d", num)
}

func main() {

    rand.Seed(time.Now().UnixNano())
    // Generate a random search tree
    randomSearchTree := avl.AVLTree[Float]{nil, 0}
    for i := 0; i < 30; i++ {
        rn := 1.0 + 99.0 * rand.Float64()
        randomSearchTree.Insert(Float(rn))
    }
    time.Sleep(3 * time.Second)
    avl.ShowTreeGraph(randomSearchTree)

    randomSearchTree.InOrderTraverse(inorderOperator)
    min := randomSearchTree.Min()
    max := randomSearchTree.Max()
    fmt.Printf("\nMinimum value in tree is %0.1f  Maximum value in tree is
                    %0.1f", *min, *max)

    /*
    start := time.Now()
    tree := avl.AVLTree[Integer]{nil, 0}
    for val := 0; val < 100_000; val++ {
        tree.Insert(Integer(val))
    }
    elapsed := time.Since(start)
```

```
    fmt.Printf("\nTime to build AVL tree with 100,000 nodes: %s.   Height of
                    tree: %d", elapsed, tree.Height())

    numbers := make([]int, 100_000)
    for i := 0; i < 100_000; i++ {
        numbers[i] = i
    }
    start = time.Now()
    sort.Ints(numbers)
    elapsed = time.Since(start)
    fmt.Printf("\nTime to sort 100_000 ints: %s", elapsed)
    */
}

/*
Time to build BST tree with 100,000 nodes: 17.054928498s
Time to build AVL tree with 100,000 nodes: 24.698786ms
Time to build AVL tree with 1_000_000 nodes: 281.799923ms
*/
```

Discussion of Main Driver Results

The graph of a 30-node AVL tree generated using Listing 10-3 is shown in Figure 10-4.

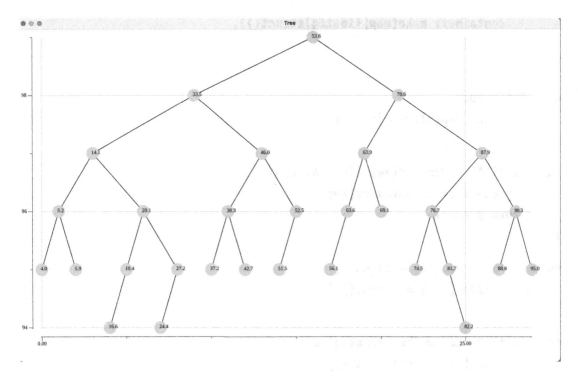

Figure 10-4. *A thirty-node AVL tree*

An AVL tree is an Ordered Set. The **Search** method allows us to determine the presence or absence of a key value in the data structure. This is a central requirement of any set. It also allows us to perform an inorder traversal that accesses the nodes from smallest to largest.

In the next section, we implement a **Set** first using an AVL tree and then using a concurrent AVL tree. We assume that the set holds floating-point values. In the next chapter, we present a more complete generic Set implementation.

10.3 Set Using Map, AVL, and Concurrent AVL

A set is typically implemented using a **map**. Listing 10-4 presents a few important methods of a set.

Listing 10-4. Set implemented using map

```
func NewSet() *Set {
    return &Set{
```

```go
        container: make(map[float64]struct{}),
    }
}

type Set struct {
    container map[float64]struct{}
}

func (c *Set) IsPresent(key float64) bool {
    _, present := c.container[key]
    return present
}

func (c *Set) Add(key float64) {
    c.container[key] = struct{}{}
}

func (c *Set) Remove(key float64) error {
    _, present := c.container[key]
    if !present {
        return fmt.Errorf("Remove Error: Item doesn't exist in set")
    }
    delete(c.container, key)
    return nil
}

func (c *Set) Size() int {
    return len(c.container)
}
```

In Listing 10-4, we assume a base type of float64 as the elements of the set. The map structure associates an empty *struct{ }* with each float64 key value. Here, we are concerned only with the key in the key-value pair in the map.

A **map** is known to produce high speed access to its members. We wish to compare the performance of this map implementation of set with an AVL tree. Following this, we define a concurrent avl set that constructs many AVL trees concurrently, and we compare its performance with the map and single AVL tree implementations.

Implementation of Set Using Map, AVL Tree, and Concurrent AVL Tree

Listing 10-5 presents a **floatset** package that includes the map, AVL, and concurrent AVL implementations of set. We skip the implementation details of AVL tree to save space since it has been presented earlier.

Listing 10-5. Package floatset

```go
package floatset

import (
    "fmt"
    "sort"
    "sync"
)

const (
    Concurrent = 32
)

var max [Concurrent]float64 // Holds the maximum value in each AVL tree

func NewSet() *Set { // Creates a new Set
    return &Set{
        container: make(map[float64]struct{}),
    }
}

type Set struct {
    container map[float64]struct{}
}

func (c *Set) IsPresent(key float64) bool {
    _, present := c.container[key]
    return present
}

func (c *Set) Add(key float64) {
```

```go
    c.container[key] = struct{}{}
}

func (c *Set) Remove(key float64) error {
    _, present := c.container[key]
    if !present {
        return fmt.Errorf("Remove Error: Item doesn't exist in set")
    }
    delete(c.container, key)
    return nil
}

func (c *Set) Size() int {
    return len(c.container)
}

// Skip AVL tree details


var concurrrentSet [Concurrent]AVLTree // Slice of AVL trees

func BuildConcurrentSet(dataSet []float64) {
    // Use concurrent processing to construct concurrent AVL trees
    var wg sync.WaitGroup
    sort.Float64s(dataSet)
    segment := len(dataSet) / Concurrent
    for treeNumber := 0; treeNumber < Concurrent; treeNumber++ {
        wg.Add(1)
        go func(num int) {
            defer wg.Done()
            startVal := segment * num
            for j := startVal; j < startVal+segment; j++ {
                concurrrentSet[num].Insert(dataSet[j])
            }
            max[num] = dataSet[startVal+segment-1]
        }(treeNumber)
    }
    wg.Wait()
```

```
}

func IsPresent(val float64) bool {
    // Determine which AVL tree val is in
    treeNumber := 0
    for ; treeNumber < len(max); treeNumber++ {
        if val <= max[treeNumber] {
            break
        }
    }
    return concurrrentSet[treeNumber].Search(val)
}
```

Explanation of Concurrent AVL Set

The constant **Concurrent** (in this case, 32) defines the number of AVL trees that we build concurrently. The variable **concurrentSet** holds an array of **AVLTree**.

First, we sort the incoming **dataSet** slice. We compute the number of nodes in each AVL, segment, by dividing the length of the **dataSet** with the number of concurrent trees.

In a loop that iterates over tree number, we invoke goroutines, each one inserting the sorted values from the incoming **dataSet** slice. The wait group assures that each concurrently constructed AVL tree is complete before we exit this function.

The global **max** array stores the maximum value in each of the AVL trees. There is no conflict among goroutines assigning to **max** since the index in **max** is unique to each goroutine (the tree number sent in).

Function **IsPresent** first determines which AVL tree the incoming **val** belongs to by comparing its value to the maximum values of each AVL tree stored in the **max** array. Once determined, the function returns the result of invoking the **Search** method on the correct tree number.

Comparing the Three Set Implementations

Listing 10-6 is a driver program that performs the experiment of comparing set construction time and most importantly the time for determining whether a value is present. To do this, we access every element in the data set and determine whether it is present in the set type we are timing.

Listing 10-6. Comparing the performance of three set types

```go
package main

import (
    "fmt"
    "math/rand"
    "time"
    "example.com/floatset"
)

const (
    size = 1_000_000
)

var dataSet []float64

func main() {
    mySet := floatset.NewSet()

    dataSet = make([]float64, size)
    for i := 0; i < size; i++ {
        dataSet[i] = 100.0 * rand.Float64()
    }
    // Time construction of Set
    start := time.Now()
    for i := 0; i < size; i++ {
        mySet.Add(dataSet[i])
    }
    elapsed := time.Since(start)
    fmt.Printf("\nTime to build Set with %d numbers: %s", size, elapsed)

    // Time to test the presence of all numbers in dataSet
    start = time.Now()
    for i := 0; i < len(dataSet); i++ {
        if !mySet.IsPresent(dataSet[i]) {
            fmt.Println("%f not present", dataSet[i])
        }
    }
```

```
elapsed = time.Since(start)
fmt.Printf("\nTime to test the presence of all numbers in Set: %s",
            elapsed)

avlSet := floatset.AVLTree{nil, 0}
// Time construction of avlSet
start = time.Now()
for i := 0; i < size; i++ {
    avlSet.Insert(dataSet[i])
}
elapsed = time.Since(start)
fmt.Printf("\n\nTime to build avlSet with %d numbers: %s", size,
elapsed)

// Time to test the presence of all numbers in avlSet
start = time.Now()
for i := 0; i < len(dataSet); i++ {
    if !mySet.IsPresent(dataSet[i]) {
        fmt.Println("%f not present", dataSet[0])
    }
}
elapsed = time.Since(start)
fmt.Printf("\nTime to test the presence of all numbers in avlSet: %s",
            elapsed)

// Use concurrent processing to construct concurrent avl trees
start = time.Now()
floatset.BuildConcurrentSet(dataSet)
elapsed = time.Since(start)
fmt.Printf("\n\nTime to build concurrent (%d) avlSet with %d numbers:
%s", floatset.Concurrent, size, elapsed)

// Test every number in dataSet against the concurrent set
start = time.Now()
for i := 0; i < len(dataSet); i++ {
    if !floatset.IsPresent(dataSet[i]) {
        fmt.Println("%f not present", dataSet[i])
```

```
        }
    }
    elapsed = time.Since(start)
    fmt.Printf("\nTime to test the presence of all numbers in concurrent
(%d) avlSet: %s", floatset.Concurrent, elapsed)
}

/*
On iMac Pro with 32G Ram and 3.2 GHz 8-Core Intel Xeon W
Time to build Set with 1000000 numbers: 184.442966ms
Time to test the presence of all numbers in Set: 105.600217ms

Time to build avlSet with 1000000 numbers: 819.517251ms
Time to test the presence of all numbers in avlSet: 103.422116ms

Time to build concurrent (32) avlSet with 1000000 numbers: 184.681628ms
Time to test the presence of all numbers in concurrent (32) avlSet:
66.183935ms

On iMac Pro Apple M1 Max with 32G Ram
Time to build Set with 1000000 numbers: 90.186209ms
Time to test the presence of all numbers in Set: 44.667542ms

Time to build avlSet with 1000000 numbers: 421.970625ms
Time to test the presence of all numbers in avlSet: 39.154042ms

Time to build concurrent (32) avlSet with 1000000 numbers: 172.478583ms
Time to test the presence of all numbers in concurrent (32) avlSet:
47.972875ms
*/
```

Discussion of Results

The program was run on two computers, and the results are surprising.

On a 2017 iMac Pro with 32G of RAM and a 3.2-GHz 8-Core Intel Xeon W processor, the ***concurrentAVLSet*** turns in the fastest ***isPresent*** performance, faster than a single AVL tree and over twice as fast as the map implementation of set.

On a MacBook Pro with 32G of unified RAM and an Apple M1 Max chip with 10-core CPU and 32-core GPU, the ***concurrentAVLSet*** turns in the slowest performance, and the single AVL tree turns in the fastest performance.

It must be noted that all the set implementations on the Apple M1 Max computer are significantly faster than their corresponding execution times on the Intel Xeon W computer.

It is therefore not clear whether the use of go-routines and concurrent processing in populating 32 AVL trees with the input data provides a meaningful benefit since the results are processor dependent.

10.4 Summary

We presented the properties of an AVL tree. The operations of **Insert** and **Delete** preserve the AVL properties. We outlined the logic for performing these operations. Then we presented a package that includes these operations and examined the performance associated with constructing and searching an AVL tree. Finally, we presented three different implementations of a Set using a map, an AVL tree, and a concurrent AVL tree and compared their performance.

In the next chapter, we focus on hash functions and hash tables along with several important applications.

CHAPTER 11

Heap Trees

The previous chapter presented AVL trees. These trees are extremely useful when many fast lookups are needed.

In this chapter, we present another important tree structure, **Heap**. A heap tree is another balanced tree type with the largest item in the tree always in the root of the tree. We use a heap tree to implement an efficient sorting algorithm.

In the next section, we define heap tree and illustrate heap tree construction.

11.1 Heap Tree Construction

A heap is a complete binary tree such that **each node has a value greater than its two children**. The largest value in a heap tree will always be in the root node. A complete tree has leaf nodes filled from left to right, all at the deepest level in the tree.

Consider the heap tree shown in the following. Each node has a value greater than its two children.

We wish to insert a new node with the value 90. See Figure 11-1.

© Richard Wiener 2022
R. Wiener, *Generic Data Structures and Algorithms in Go*, https://doi.org/10.1007/978-1-4842-8191-8_11

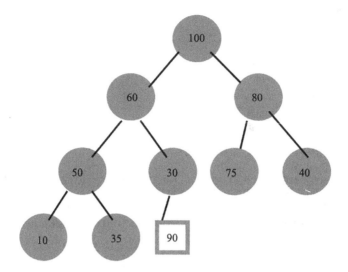

Figure 11-1. *Insertion of 90*

We fill the leaf nodes from left to right, so node 90 needs to be the left node of 30. But 90 is larger than its parent 30. So we exchange the two nodes. See Figure 11-2.

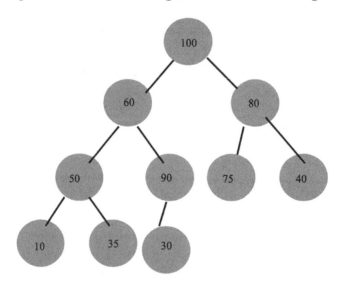

Figure 11-2. *Insertion continued*

But 90 is larger than its parent 60, so we do another exchange producing the new heap tree that contains 90. See Figure 11-3.

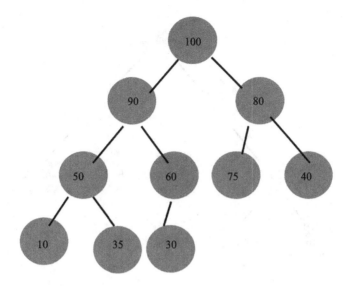

Figure 11-3. *Result after insertion*

In the next section, we show how to perform deletion from a heap tree.

11.2 Deletion from a Heap Tree

We can only delete the value in the root node of a heap tree. To delete the root node value 100, we replace the value in the root node with the value in the rightmost node on the lowest level of the tree, 30 in this case. Then we compare the new root value with the values of its two children, swapping with the larger of the children. We continue this sift-down process until there are no further nodes to swap. So 30 gets swapped with 90 (the largest of the children, 90 and 80); then 30 gets swapped with 60 (the larger of the two children, 50 and 60). This leads to the new heap tree shown in Figure 11-4.

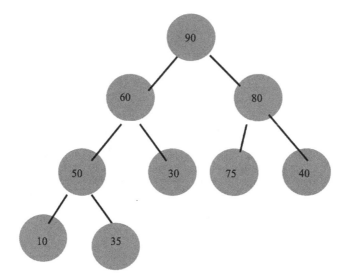

Figure 11-4. *Result after deletion*

In the next section, we examine the implementation details for building a generic heap tree from a slice of items and inserting a new item.

11.3 Implementation of a Heap Tree
Logic for Building a Heap Tree

The logic for building and inserting items in a heap tree flows from the following relationship between the **index** of an item in a slice and the location of that item in a heap tree. Suppose we have an item at a specified **index**.

- Its parent is at location **index / 2** if index is odd and at location **index / 2 – 1** if index is even.

- Its left child is at index **2 * index + 1**.

- Its right child is at index **2 * index + 2**.

Consider the slice **[90, 60, 80, 50, 30, 75, 40, 10, 35]** that corresponds to the preceding heap tree. The slice values relate to the values in the heap tree by traversing the values from left to right at each succeeding level in the tree.

Consider the node with value 50 at index 3 in the slice.

The parent is at index 3 / 2, which equals 1. This corresponds to the node with value 60. The two children are at index values 2 * 3 + 1 and 2 * 3 + 2 or indices 7 and 8. This corresponds to the nodes with values 10 and 35.

Package Heap

Listing 11-1 presents a package for a generic heap, and Listing 11-2 shows a main driver program to test and exercise the methods of package heap.

Listing 11-1. Package heap

```
package heap

type Ordered interface {
    ~float64 | ~int | ~string
}

type Heap[T Ordered] struct {
    Items []T
}

// Methods
func (heap *Heap[T]) Swap(index1, index2 int) {
    heap.Items[index1], heap.Items[index2] =
            heap.Items[index2], heap.Items[index1]
}

func NewHeap[T Ordered](input []T) *Heap[T] {
    heap := &Heap[T]{}
    for i := 0; i < len(input); i++ {
        heap.Insert(input[i])
    }
    return heap
}

func (heap *Heap[T]) Insert(value T) {
    heap.Items = append(heap.Items, value)
    heap.buildHeap(len(heap.Items) - 1)
}
```

```go
func (heap *Heap[T]) Remove() {
    // Can only remove Items[0], the largest value
    heap.Items[0] = heap.Items[len(heap.Items)-1]
    heap.Items = heap.Items[:(len(heap.Items) - 1)]
    heap.rebuildHeap(0)
}

func (heap *Heap[T]) Largest() T {
    return heap.Items[0]
}

func (heap *Heap[T]) buildHeap(index int) {
    var parent int
    if index > 0 {
        parent = (index - 1) / 2
        if heap.Items[index] > heap.Items[parent] {
            heap.Swap(index, parent)
        }
        heap.buildHeap(parent)
    }
}

func (heap *Heap[T]) rebuildHeap(index int) {
    length := len(heap.Items)
    if (2 * index + 1) < length {
        left := 2*index + 1
        right := 2*index + 2
        largest := index

        if left < length && right < length &&
            heap.Items[left] >= heap.Items[right] &&
                heap.Items[index] < heap.Items[left] {
            largest = left
        } else if right < length &&
            heap.Items[right] >= heap.Items[left] &&
                heap.Items[index] < heap.Items[right]{
            largest = right
        } else if left < length && right >= length &&
```

```
                    heap.Items[index] < heap.Items[left] {
                largest = left
            }
            if index != largest {
                heap.Swap(index, largest)
                heap.rebuildHeap(largest)
            }
        }
    }
}
```

Listing 11-2. Main driver for heap

package main

```
import (
    "fmt"
    "example.com/heap"
)

func main() {
    slice1:= []int{100, 60, 80, 50, 30, 75, 40, 10, 35}
    heap1 := heap.NewHeap[int](slice1)
    heap1.Insert(90)
    fmt.Println("heap1 after inserting 90")
    fmt.Println(heap1.Items)
    fmt.Println("Largest item in heap: ", heap1.Largest())

    heap1.Remove()
    fmt.Println("Removing largest item from heap
                yielding the heap: ")
    fmt.Println(heap1.Items)
    fmt.Println("Largest item in heap: ", heap1.Largest())

    slice2:= []int{10, 35, 100, 80, 30, 75, 40, 50, 60}
    heap2 := heap.NewHeap[int](slice2)
    heap2.Insert(90)
    fmt.Println("heap2 with rearranged slice2 after inserting 90")
    fmt.Println(heap2.Items)
}
```

```
/* Output
heap1 after inserting 90
[100 90 80 50 60 75 40 10 35 30]
Largest item in heap:   100
Removing largest item from heap yielding the heap:
[90 60 80 50 30 75 40 10 35]
Largest item in heap:   90
heap2 with rearranged slice2 after inserting 90
[100 90 75 60 80 35 40 10 50 30]
*/
```

Explanation of Package heap

The generic **Heap** structure is given by a struct containing a slice of generic ordered type **T**.

```
type Heap[T Ordered] struct {
    Items []T
}
```

We focus on the function **NewHeap** and on the methods **Insert** and **Remove**. The other methods are much simpler and do not need explanation.

To build a heap from a slice of some ordered type **T**, we perform

```
func NewHeap[T Ordered](input []T) *Heap[T] {
    heap := &Heap[T]{}
    for i := 0; i < len(input); i++ {
        heap.Insert(input[i])
    }
    return heap
}
```

The first line of code defines a **heap** as the address (since we are returning a pointer to a Heap) of **Heap** with an empty slice of **Items**.

A for-loop follows that invokes the **Insert** method on each item in the **input** slice.

The *Insert* method, given as

```
func (heap *Heap[T]) Insert(value T) {
    heap.Items = append(heap.Items, value)
    heap.buildHeap(len(heap.Items) - 1)
}
```

appends the input *value* to *the heap.Items* slice. It then invokes the private method *buildHeap*.

This private method *buildHeap* directly follows the example shown in Section 11.1 and works upward from the bottom of the tree doing swaps when necessary to produce a heap.

The *Remove* method, given as

```
func (heap *Heap[T]) Remove() {
    // Can only remove Items[0], the largest value
    heap.Items[0] = heap.Items[len(heap.Items)-1]
    heap.Items = heap.Items[:(len(heap.Items) - 1)]
    heap.rebuildHeap(0)
}
```

assigns the item in the lowest rightmost position to index 0 in the *heap.Items* slice. It then reassigns this slice to exclude this rightmost item. The heap structure is temporarily broken by placing the deepest, rightmost value in the root. A private method *rebuildHeap* is invoked, which restores the heap property.

The method *rebuildHeap* is closely reasoned and requires care in understanding how it works. At each level of recursion, the item at *index* is initially assumed to be the largest. The values at index *left* and index *right* (or just *left* if *right* is out of range) are compared. If the value at *index* is less than the larger of the children, *largest* is set to the index of the larger child. Then a swap of values between value at *index* and *largest* is made, and a recursive call to *rebuildHeap* is made with parameter *largest* sent into *rebuildHeap*. Upon the completion of this method, the heap structure is restored.

Since a heap is close to perfectly balanced, its height is related to the number of nodes with a logarithmic relationship, $height = \log_2 n$, where n is the number of nodes. Therefore, the methods *buildHeap* and *rebuildHeap* have complexity $O(\log_2 n)$.

In the main driver program, a second heap, *heap2*, is constructed using the same input integers but arranged in a different order. The resulting tree is indeed a heap but with a slightly different sequence of values in the slice.

In the next section, we examine an important application of a heap tree – a sorting algorithm, heap sort.

11.4 Heap Sort

The heap tree provides the basis for a sorting algorithm. It works as follows:

Build a heap from the initial list to be sorted. Extract the largest from the root and append it to the result list (initialized to empty). Apply the ***Remove*** method to the heap. Continue this process until the heap is shrunk to empty.

This process will produce a slice sorted from largest to smallest. We can produce output in ascending order by reversing the sequence in the slice produced previously.

The details of heap sort are presented in Listing 11-3.

Listing 11-3. Heap sort

```go
package main

import (
    "example.com/heap"
    "fmt"
    "math/rand"
    "time"
)

type Ordered interface {
    ~float64 | ~int | ~string
}

func heapSort[T Ordered](input []T) []T {
    heap1 := heap.NewHeap[T](input)
    descending := []T{}
    for {
        if len(heap1.Items) > 0 {
            descending = append(descending, heap1.Largest())
```

```
                heap1.Remove()
        } else {
            break
        }
    }
    ascending := []T{}
    for i := len(descending) - 1; i >= 0; i-- {
        ascending = append(ascending, descending[i])
    }
    return ascending
}

const size = 50_000_000

func IsSorted[T Ordered](data []T) bool {
    for i := 1; i < len(data); i++ {
        if data[i] < data[i-1] {
            return false
        }
    }
    return true
}

func main() {
    slice := []float64{0.0, 2.7, -3.3, 9.6, -13.8, 26.0, 4.9, 2.6,
    5.1, 1.1}
    sorted := heapSort[float64](slice)
    fmt.Println("After heapSort on slice: ", sorted)

    data := make([]float64, size)
    for i := 0; i < size; i++ {
        data[i] = 100.0 * rand.Float64()
    }
    start := time.Now()
    largeSorted := heapSort[float64](data)
    elapsed := time.Since(start)
    fmt.Println("Time for heapSort of 50 million floats: ", elapsed)
```

```
    if !IsSorted[float64](largeSorted) {
        fmt.Println("largeSorted is not sorted.")
    }
}
}
/* Output
Elapsed time for regular quicksort = 5.382400384s   (from Chapter 1)
Elapsed time for concurrent quicksort = 710.431619ms (from Chapter 1)

After heapSort on slice:  [-13.8 -3.3 0 1.1 2.6 2.7 4.9 5.1 9.6 26]
Time for heapSort of 50 million floats:  23.978801647s
*/
```

Discussion of heapsort Results

The complexity of heapsort is **O(nlog$_2$n)** since the complexity of ***buildHeap*** and ***rebuildHeap*** is **log$_2$n,** and we do this n times.

Comparing the time to sort 50 million floating-point numbers with quicksort or concurrent quicksort, we see that heapSort is about four times slower than quicksort.

In the next section, we examine another application of Heap, a priority queue.

11.5 Heap Application: Priority Queue

A heap provides a natural model for a priority queue. Each item is assumed to encapsulate a priority. For example, if we insert string values into the priority queue, we assume that the larger the string in a lexical sense, the higher its priority. So the string "Zachary" has a higher priority than the string "Robert".

Listing 11-4 shows an implementation of priority queue using heaps.

Listing 11-4. Priority queue using heap

```
package main

import (
    "example.com/heap"
    "fmt"
)
```

```go
type Ordered interface {
    ~float64 | ~int | ~string
}

type PriorityQueue[T Ordered] struct {
    infoHeap heap.Heap[T]
}

// Methods
func (queue *PriorityQueue[T]) Push(item T) {
    queue.infoHeap.Insert(item)
}

func (queue *PriorityQueue[T]) Pop() T {
    returnValue := queue.infoHeap.Largest()
    queue.infoHeap.Remove()
    return returnValue
}

func main() {
    myQueue := PriorityQueue[string]{}
    myQueue.Push("Helen")
    myQueue.Push("Apollo")
    myQueue.Push("Richard")
    myQueue.Push("Barbara")
    fmt.Println(myQueue)
    myQueue.Pop()
    fmt.Println(myQueue)
    myQueue.Push("Arlene")
    fmt.Println(myQueue)
    myQueue.Pop()
    myQueue.Pop()
    fmt.Println(myQueue)
}
/* Output
{{[Richard Barbara Helen Apollo]}}
{{[Helen Barbara Apollo]}}
```

```
{{[Helen Barbara Apollo Arlene]}}
{{[Arlene Apollo]}}
*/
```

11.6 Summary

In this chapter, we defined a heap structure and presented an implementation. Building a heap from a slice of items guarantees that the largest item is in the root node. Every item in a heap is larger than its left and right child items. We used a heap to implement an efficient sorting algorithm. We also used a heap to implement a priority queue.

In the next chapter, we introduce and implement red-black trees.

CHAPTER 12

Red-Black Trees

In the previous chapter, we presented heap trees. These are close to fully balanced trees in which the largest item is always found in the root node and each node has a value greater than its children.

In this chapter, we present another balanced tree structure, the red-black tree. Like the AVL tree presented in Chapter 10, the red-black tree data structure is aimed at efficient insertion, deletion, and searching of items stored in the tree.

In the next section, we introduce red-black trees.

12.1 Red-Black Trees

An interesting and important balanced binary search tree is the red-black tree. Rudolf Bayer invented this tree structure in 1972, ten years after the AVL tree was invented.

Red-black trees, like AVL trees, are self-balancing. After an insertion or deletion, the resulting tree is a red-black tree. Like AVL trees, the computational complexity for insertion, deletion, or search is **O(log$_2$n)**.

Insertion and deletion for red-black trees generally involve fewer rotational corrections, but the resulting tree is less balanced than an AVL tree. In applications that expect many insertions and deletions and fewer searches, red-black trees may be preferable to AVL trees.

Because of the complexity of red-black trees, we limit ourselves in this chapter to implementing insertion into a red-black tree. The interested reader will find an implementation for deletion in Chapter 13 (page 545) of my book, Modern Software Development Using C#.Net, Thompson, 2006.

Definition of Red-Black Tree

A binary search tree is a red-black tree if

© Richard Wiener 2022
R. Wiener, *Generic Data Structures and Algorithms in Go*, https://doi.org/10.1007/978-1-4842-8191-8_12

1. Every node is assigned a color of red or black.

2. The root node is always black.

3. The children of a red node are black.

4. Every path from the root node to a leaf node contains the same number of black nodes.

Example of Red-Black Tree

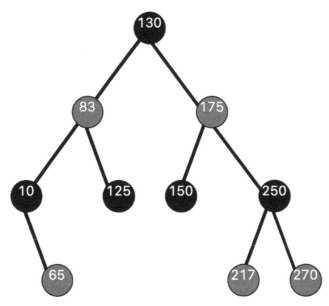

In this ten-node red-black tree, every path from the root to a leaf node contains exactly two black nodes.

Some terminologies we will use include **parent**, **grandparent,** and **uncle**.

As an example, the parent of node 217 is 250. The uncle of 217 is 150 (sibling of parent). The grandparent of 217 is 175.

In the next section, we discuss the logic of inserting an item into a red-black tree. We "walk" through an example in detail to illustrate the process.

12.2 Insertion Process

We discuss the logic of insertion with a series of examples.

The first step for insertion is to do an ordinary search-tree insertion.

The new node added to the tree is always colored red. Our goal is to keep the number of black nodes between root and all leaf nodes constant.

If the new inserted node has a red parent, this violates condition 3 in the preceding text, which requires the child of a red node to be black. We then must take corrective action.

The first case we consider is when the **parent of the node inserted is red** and **the uncle of the node inserted exists and is red**. Consider the following tree after inserting 25. The uncle of 25 is 150 and is red.

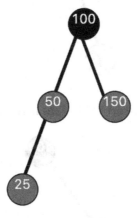

We perform a correction by modifying the color of the **parent** (change red to black) and **uncle** (change red to black) and **grandparent** if it is not the root (which it is in this case). The corrected tree is shown in the following. If 100 was not the root, we might have to continue the search for violations up the tree after changing node 100 to red.

The result of performing color modification is shown as follows:

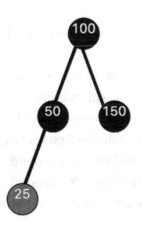

The next case we consider is when the **parent of the node inserted is red** and **the uncle is black or does not exist**. There are four cases to consider.

In the first case, we insert 25. Parent is red and uncle does not exist.

In the second case, we again insert 25. Parent is red and uncle does not exist.

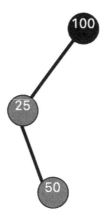

The other two cases are symmetric with respect to the root node (are on the right side of the root).

The corrective action we take involves tree rotations as follows:

We take an inorder traversal of the subtree starting at the grandparent and label the nodes **first**, **second**, and **third** in the traversal; then the second node will always be the new root of the subtree and its left child the **first** and right child the **third**.

In case 1, the traversal produces **first = 25, second = 50**, and **third = 100.**

In case 2, the traversal produces **first = 25, second = 50**, and **third = 100.**

We recolor the new subroot black, and its two children remain red.

This produces the corrected tree.

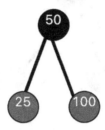

In case 1, we perform a **right rotate on node 100**. In case 2, we perform a **left rotate on node 25** (producing case 1) and then a **right rotate on node 100**. Cases 3 and 4 follow a symmetric pattern.

Detailed Walk-Through of Many Insertions

To solidify our understanding of insertion, we construct a red-black tree, step by step, by inserting the sequence of values: 10, 20, 4, 15, 17, 40, 50, 60, 70, 35, 38, 18, 19, 45, 30, 25. We show the work for some of the insertions and leave the rest as an exercise.

After inserting 10, 20, and 4, we have

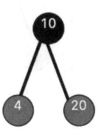

After inserting 15, we have

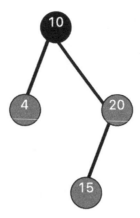

Since the parent of 15 is red and uncle is red, we do recoloring to produce

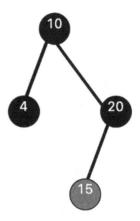

After inserting 17, we get

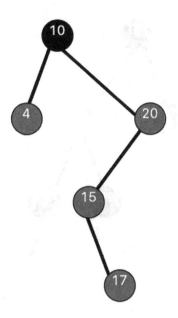

But this needs correction. Since the parent of 17 is red and uncle does not exist, we perform rotational corrections (left on 15 and right on 20) and recoloring to get

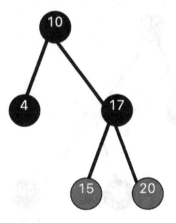

We next insert 40. We show only the result after reconfiguring (recoloring case) because the parent and uncle are red.

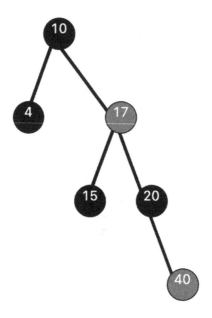

We next insert 50. This is a case 4 requiring one left rotational correction on 20 producing

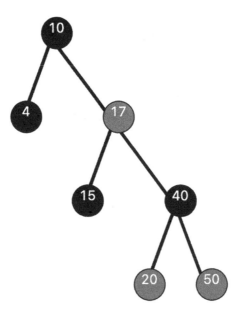

We next insert 60. Because of the red parent and red uncle, this requires only recoloring. The result is

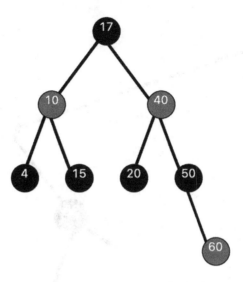

We are halfway there! As an exercise, please continue the insertions and show that the final red-black tree is

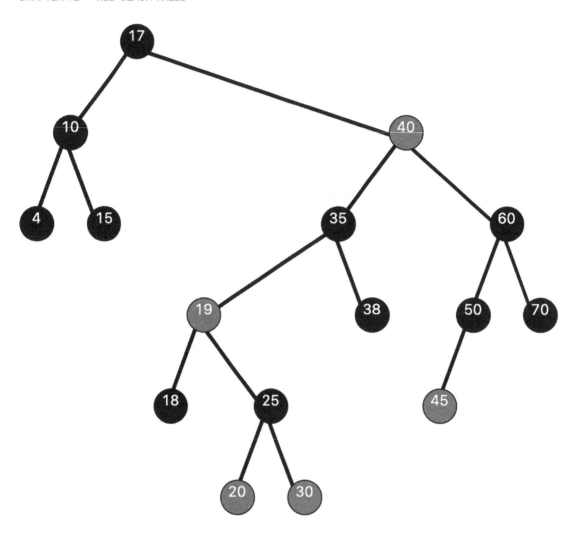

A careful inspection of this tree shows that the number of black nodes from root 17 to every leaf is exactly 3. Every red node has only black children.

This tree is clearly less balanced than an AVL tree (the maximum depth on the right side of the root is 5 and the maximum depth on the left side of the root is 2).

In the next section, we present an implementation of **Insertion** into a red-black tree. The details are complex because of the many special cases.

12.3 Implementation of Red-Black Tree

The implementation details for insertion into a red-black tree are daunting. This is because of the number of possible rotational or color corrections that are potentially possible based on the logic discussed and illustrated in Section 12.2.

The best strategy for unraveling the logic in the implementation presented in Listing 12-1 is to "walk" as far as you can, step by step, through the example presented in Section 12.2.

A few small changes to the display tree function, defined and discussed in Section 8.3, were made for drawing a red-black tree. The changes in this portion of the implementation are shown in boldface.

Listing 12-1 presents the implementation of a red-black tree, including logic for drawing the tree, but only including the *Insert* method. The tree implementation is combined with a short driver program, *main*, without creating a separate package for the tree.

Listing 12-1. Red-black tree

```go
package main

import (
    "image/color"
    "log"
    "fyne.io/fyne/v2"
    "fyne.io/fyne/v2/app"
    "fyne.io/fyne/v2/canvas"
    "fyne.io/fyne/v2/theme"
    "github.com/mitchellh/go-homedir"
    "gonum.org/v1/plot"
    "gonum.org/v1/plot/plotter"
    "gonum.org/v1/plot/vg"
    "gonum.org/v1/plot/vg/draw"
    "strconv"
)

type ordered interface {
    ~int | ~float64 | ~string
}
```

373

```
type OrderedStringer interface {
    ordered
    String() string
}
type Node[T OrderedStringer] struct {
    value T
    red bool
    parent *Node[T]
    left *Node[T]
    right *Node[T]
}

type RedBlackTree[T OrderedStringer] struct {
    count int
    root *Node[T]
}

func NewTree[T OrderedStringer](value T) *RedBlackTree[T] {
    return &RedBlackTree[T]{1, &Node[T]{value, false, nil, nil, nil}}
}

// Methods
func (tree *RedBlackTree[T]) Insert(value T) {
    if tree.root == nil { // Empty tree
        tree.root = &Node[T]{value, false, nil, nil, nil}
        tree.count += 1
        return
    }
    parent, nodeDirection := tree.findParent(value)
    if nodeDirection == "" {
        return
    }
    newNode := Node[T]{value, true, parent, nil, nil}

    if nodeDirection == "L" {
        parent.left = &newNode
```

```go
    } else {
        parent.right = &newNode
    }
    tree.checkReconfigure(&newNode)
    tree.count += 1
}

func (tree *RedBlackTree[T]) IsPresent(value T, node
        *Node[T]) bool {
    if node == nil {
        return false
    }
    if value < node.value {
        return tree.IsPresent(value, node.left)
    }
    if value > node.value {
        return tree.IsPresent(value, node.right)
    }
    return true
}

func (tree *RedBlackTree[T]) findParent(value T)
                (*Node[T], string) {
    return search(value, tree.root)
}

func (tree *RedBlackTree[T]) checkReconfigure(node *Node[T]) {
    var nodeDirection, parentDirection, rotation string
    var uncle *Node[T]

    parent := node.parent
    value :=  node.value
    if parent == nil || parent.parent == nil ||
                node.red == false || parent.red == false {
        return
    }
    grandfather := parent.parent
```

```
    if value < parent.value {
        nodeDirection = "L"
    } else {
        nodeDirection = "R"
    }
    if grandfather.value > parent.value {
        parentDirection = "L"
    } else {
        parentDirection = "R"
    }
    if parentDirection == "L" {
        uncle = grandfather.right
    } else {
        uncle = grandfather.left
    }
    rotation = nodeDirection + parentDirection
    if uncle == nil || uncle.red == false {
        if rotation == "LL" {
            tree.rightRotate(node, parent, grandfather, true)
        } else if rotation == "RR" {
            tree.leftRotate(node, parent, grandfather, true)
        } else if rotation == "LR" {
            tree.rightRotate(nil, node, parent, false)
            tree.leftRotate(parent, node, grandfather, true)
            node, parent = parent, node

        } else if rotation == "RL" {
            tree.leftRotate(nil, node, parent, false)
            tree.rightRotate(parent, node, grandfather, true)
        }
    } else {
        tree.modifyColor(grandfather)
    }
}

func (tree *RedBlackTree[T]) leftRotate(node, parent, grandfather
*Node[T],  modifyColor bool) {
```

```
    greatgrandfather := grandfather.parent
    tree.updateParent(parent, grandfather, greatgrandfather)
    oldLeft := parent.left
    parent.left = grandfather
    grandfather.parent = parent
    grandfather.right  = oldLeft
    if oldLeft != nil {
        oldLeft.parent = grandfather
    }
    if modifyColor == true {
        parent.red = false
        node.red = true
        grandfather.red = true
    }
}

func (tree *RedBlackTree[T]) rightRotate(node, parent,
        grandfather *Node[T],  modifyColor bool) {
    greatgrandfather := grandfather.parent
    tree.updateParent(parent, grandfather,
                      greatgrandfather)
    oldRight := parent.right
    parent.right = grandfather
    grandfather.parent = parent
    grandfather.left = oldRight
    if oldRight != nil {
        oldRight.parent = grandfather
    }
    if modifyColor == true {
        parent.red = false
        node.red = true
        grandfather.red = true
    }
}
```

```
func (tree *RedBlackTree[T]) modifyColor(grandfather
        *Node[T]) {
    grandfather.right.red = false
    grandfather.left.red = false
    if grandfather != tree.root {
        grandfather.red = true
    }
    tree.checkReconfigure(grandfather)
}

func (tree *RedBlackTree[T]) updateParent(node,
            parentOldChild, newParent *Node[T]) {
    node.parent = newParent
    if newParent != nil {
        if newParent.value > parentOldChild.value {
            newParent.left = node
        } else {
            newParent.right = node
        }
    } else {
        tree.root = node
    }
}

func search[T OrderedStringer](value T, node *Node[T])
                (*Node[T], string) {
    if value == node.value {
        return nil, ""
    } else if value > node.value {
        if node.right == nil {
            return node, "R"
        }
        return search(value, node.right)
    } else if value < node.value {
        if node.left == nil {
            return node, "L"
        }
```

```go
        return search(value, node.left)
    }
    return nil, ""
}
// Logic for drawing tree
type NodePair struct {
    Val1, Val2 string
}

type NodePos struct {
    Val string
    Red bool
    YPos int
    XPos int
}

var data []NodePos
var endPoints []NodePair // Used to plot lines

func PrepareDrawTree[T OrderedStringer](tree RedBlackTree[T]) {
    prepareToDraw(tree)
}

func FindXY(val interface{}) (int, int) {
    for i := 0; i < len(data); i++ {
        if data[i].Val == val {
            return data[i].XPos, data[i].YPos
        }
    }
    return -1, -1
}

func FindX(val interface{}) int {
    for i := 0; i < len(data); i++ {
        if data[i].Val == val {
            return i
        }
    }
}
```

```
        return -1
}
func SetXValues() {
    for index := 0; index < len(data); index++ {
        xValue := FindX(data[index].Val)
        data[index].XPos = xValue
    }
}

func prepareToDraw[T OrderedStringer](tree RedBlackTree[T]) {
    inorderLevel(tree.root, 1)
    SetXValues()
    getEndPoints(tree.root, nil)
}

func inorderLevel[T OrderedStringer](node *Node[T], level int) {
    if node != nil {
        inorderLevel(node.left, level + 1)
        data = append(data,
            NodePos{node.value.String(), node.red,
                    100 - level, -1})
        inorderLevel(node.right, level + 1)
    }
}

func getEndPoints[T OrderedStringer](node *Node[T], parent *Node[T]) {
    if node != nil {
        if parent != nil {
            endPoints = append(endPoints,
                NodePair{node.value.String(),
                    parent.value.String()})
        }
        getEndPoints(node.left, node)
        getEndPoints(node.right, node)
    }
}

var path string
```

```go
func DrawGraph(a fyne.App, w fyne.Window) {
    image := canvas.NewImageFromResource(theme.FyneLogo())
    image = canvas.NewImageFromFile(path + "tree.png")
    image.FillMode = canvas.ImageFillOriginal
    w.SetContent(image)
    w.Close()
    w.Show()
}

func ShowTreeGraph[T OrderedStringer](myTree RedBlackTree[T]) {
    PrepareDrawTree(myTree)
    myApp := app.New()
    myWindow := myApp.NewWindow("Tree")
    myWindow.Resize(fyne.NewSize(1000, 600))
    path, _ := homedir.Dir()
    path += "/Desktop//"

    nodePts := make(plotter.XYs, myTree.count)
    for i := 0; i < len(data); i++ {
        nodePts[i].Y = float64(data[i].YPos)
        nodePts[i].X = float64(data[i].XPos)
    }
    nodePtsData := nodePts
    p := plot.New()
    p.Add(plotter.NewGrid())
    nodePoints, err := plotter.NewScatter(nodePtsData)
    if err != nil {
        log.Panic(err)
    }
    nodePoints.Shape = draw.CircleGlyph{}
    nodePoints.Color = color.RGBA{R: 255, G: 255, B:
                            250, A: 255} // White fill
    nodePoints.Radius = vg.Points(12)
    // Plot lines
    for index := 0; index < len(endPoints); index++ {
        val1 := endPoints[index].Val1
        x1, y1 := FindXY(val1)
```

```
    val2 := endPoints[index].Val2
    x2, y2 := FindXY(val2)
    pts := plotter.XYs{{X: float64(x1), Y:
        float64(y1)},{X: float64(x2), Y: float64(y2)}}
    line, err := plotter.NewLine(pts)
    if err != nil {
        log.Panic(err)
    }
    scatter, err := plotter.NewScatter(pts)
    if err != nil {
        log.Panic(err)
    }
    p.Add(line, scatter)
}
p.Add(nodePoints)

// Add Labels
for index := 0; index < len(data); index++ {
    x := float64(data[index].XPos) - 0.10
    y := float64(data[index].YPos) - 0.02
    str := data[index].Val
    if data[index].Red == true {
        str += "(RED)"
    } else {
        str += "(BLACK)"
    }
    label, err :=
        plotter.NewLabels(plotter.XYLabels {
        XYs: []plotter.XY {
            {X: x ,Y: y},
        },
        Labels: []string{str},
        },)
```

```
        if err != nil {
            log.Fatalf("could not creates labels
                    plotter: %+v", err)
        }
        p.Add(label)
    }

    path, _ = homedir.Dir()
    path += "/Desktop/GoDS/"
    err = p.Save(1000, 600, "tree.png")
    if err != nil {
        log.Panic(err)
    }

    DrawGraph(myApp, myWindow)


    myWindow.ShowAndRun()
}

// Make int comply with Stringer interface
type Integer int

func (i Integer) String() string {
    return strconv.Itoa(int(i))
}

func main() {
    myTree := NewTree[Integer](10)
    myTree.Insert(20)
    myTree.Insert(4)
    myTree.Insert(15)
    myTree.Insert(17)
    myTree.Insert(40)
    myTree.Insert(50)
    myTree.Insert(60)
    myTree.Insert(70)
    myTree.Insert(35)
    myTree.Insert(38)
```

```
    myTree.Insert(18)
    myTree.Insert(19)
    myTree.Insert(45)
    myTree.Insert(30)
    myTree.Insert(25)
    ShowTreeGraph(*myTree)
}
```

The output produced by main is shown in the following. This is the same as the tree constructed in Section 12.2.

The *OrderedStringer* interface was brought back into play because the display tree requires it to create the labels for each tree node.

Comparing the Performance of Red-Black Tree to AVL Tree

A benchmark test was performed to see how long it takes to construct a red-black tree from a sequence of 100,000 random integers. The same test was performed to see the time required to build an AVL tree from 100,000 random integers.

The results are interesting and the following:

Insertion time for red-black tree: **27.62615ms**

Search time for red-black tree: **16.037945ms**

Insertion time for AVL tree: **48.315163ms**

Search time for AVL tree: **3.914522ms**

Benchmark Conclusion

The red-black tree takes about **half as long to build** but takes **four times as long to search** compared to the AVL tree. The AVL tree is more balanced than the red-black tree but requires many more rotations during construction.

Since we typically build search trees for many fast lookups, the AVL is generally preferable in such cases.

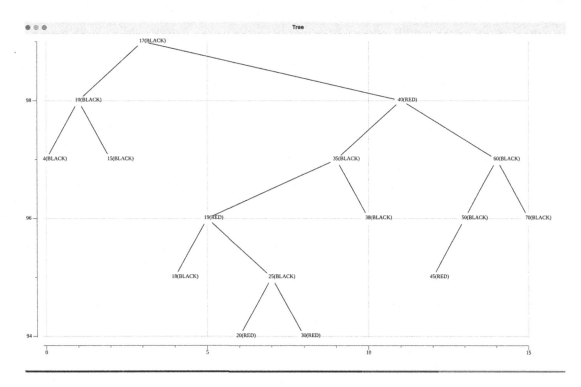

12.4 Summary

The logic for building a red-black tree was presented and illustrated. An implementation of a generic red-black tree was presented with the **Insert** method along with many supporting methods. With small modifications, the code for drawing a red-black tree was shown. The performance of a red-black tree was compared to an AVL tree. Red-black trees can be more efficiently generated but are less efficient to search than AVL trees.

In the next chapter, we introduce expression trees.

CHAPTER 13

Expression Trees

In the previous chapter, we presented red-black trees. These binary search trees provide faster insertion performance compared to AVL trees but slower search time.

In this chapter, we introduce and implement expression trees. These are used to represent and evaluate some mathematical expressions.

In the next section, we introduce expression trees.

13.1 Expression Trees

Expression trees are used to represent and evaluate mathematical expressions. Here, we limit such expressions to have operands given by a single character between "a" and "z" and operators that include "+", "-", "*", and "/".

Consider the expression "((a + b) + (c - d) / (f + g) + h)) + y / (x - z)".

An expression tree representing this mathematical expression is shown in Figure 13-1.

© Richard Wiener 2022
R. Wiener, *Generic Data Structures and Algorithms in Go*, https://doi.org/10.1007/978-1-4842-8191-8_13

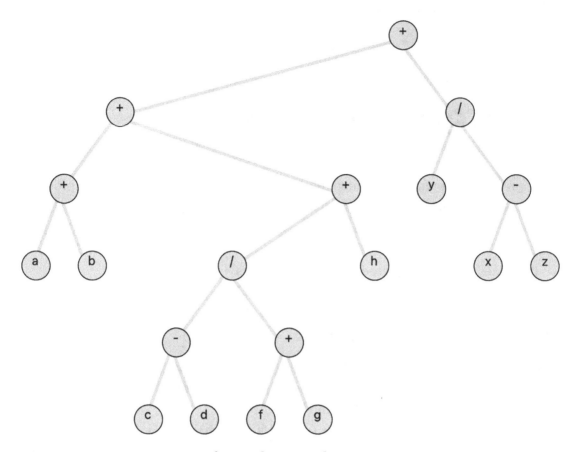

Figure 13-1. *Expression tree for mathematical expression*

The operands are contained in the leaf nodes and the operators in the interior nodes.

We interpret and obtain the mathematical expression represented by this tree by starting at the various leaf nodes and working upward toward the root node.

Starting with the leftmost leaf nodes, we have (a + b) + ...

Moving to the middle section leaf nodes, we have

$$(c - d) / (f + g) + h$$

From the rightmost leaf nodes, we have

$$y / (x - z) + ...$$

Putting the three sections together gives us the original expression.

In the next section, we present and discuss the construction of an expression tree.

13.2 Construction of an Expression Tree

The construction of an expression tree requires a layering of abstractions. We need a Stack to assist in the construction process.

We define two types, *Node* and *ExpressionTree*, as follows:

```
type Node struct {
    ch string
    left *Node
    right *Node
}

type ExpressionTree struct {
    postfix string
    root *Node
}
```

Type *Node* is the familiar binary tree node with a string, *ch*, stored in each node. This string will be either an operand or operator.

Type *ExpressionTree* contains two fields. Field *postfix* is the postfix string representation of the mathematical expression we input to build the expression tree. Field *root* is a pointer to *Node*.

Building a New Expression Tree

Function *NewTree*, presented in the following, is used to build our expression tree.

```
func NewTree(infix string) (tree *ExpressionTree) {
    infix = strings.ToLower(infix)
    tree = &ExpressionTree{"", nil}
    tree.postfix = infixpostfix(infix)
    stack := nodestack.Stack[*Node]{} // Create stack
                                       // of Node
    str := strings.Split(tree.postfix, "")
    for index := 0; index < len(str); index++ {
        if str[index] >= string('a') &&
                str[index] <= string('z') {
            node := &Node{str[index], nil, nil}
```

389

```
                stack.Push(node)
        } else if (str[index] == "+") ||
                  (str[index] == "-") ||
                  (str[index] == "*") ||
                  (str[index] == "/") {
            right := stack.Top()
            stack.Pop()
            left := stack.Top()
            stack.Pop()
            node := &Node{str[index], nil, nil}
            node.left = left
            node.right = right
            stack.Push(node)
        }
    }
    tree.root = stack.Top()
    return tree
}
```

Explanation of Function NewTree

The first four lines of code create an empty tree (the *tree* variable is used as the return variable) and an empty **Stack** of base type pointer to *Node*.

This is another example of how useful generic data structures are. Instead of having to duplicate a new stack implementation with **Node* as a base type, we can simply use the generic stack package and specify the base type as **Node*.

In a for-loop that accesses each character of the postfix string, if the character is an operand, we create a node with the character and push the node onto the stack.

If the character is one of the four possible operators, we grab the top two characters from the stack, create a node containing the operator character, and set its left and right child to the two nodes popped from the stack. Finally, we push this new node onto the stack. This is equivalent to moving upward from the leaf nodes to the root of the tree that we described in the previous section.

Function Evaluation Using Expression Tree

Method **Evaluate**, presented in the following, takes the root of an expression tree as its first parameter and a map of operand values as its second parameter and returns the value of the function (float64).

```go
func (tree *ExpressionTree) Evaluate(node *Node,
        operandValues map[string]float64) float64 {
    if node == nil {
        return 0.0
    }
    if node.left == nil && node.right == nil {
        value := operandValues[node.ch]
        return value
    }
    leftValue := tree.Evaluate(node.left, operandValues)
    rightValue := tree.Evaluate(node.right, operandValues)
    if node.ch == "+" {
        return leftValue + rightValue
    } else if node.ch == "-" {
        return leftValue - rightValue
    } else if node.ch == "*" {
        return leftValue * rightValue
    } else {
        return leftValue / rightValue
    }
}
```

Explanation of Method Evaluate

If the expression tree node is a leaf node, we assign and return **value** by accessing the **operandValues** map.

Otherwise, we assign **leftValue** and **rightValue** by recursively invoking **Evaluate** sending in **node.left** and **node.right**, along with the **operandValues** map.

Then, based on the operator contained in **node**, we combine **leftValue** and **rightValue** accordingly.

In Listing 13-1, we present the full implementation of expression tree construction and evaluation along with a **main** driver.

Listing 13-1. Expression tree

```go
package main

import (
    "fmt"
    "example.com/nodestack"
    "strings"
)

type Node struct {
    ch string
    left *Node
    right *Node
}

type ExpressionTree struct {
    postfix string
    root *Node
}

func NewTree(infix string) (tree *ExpressionTree) {
    infix = strings.ToLower(infix)
    tree = &ExpressionTree{"", nil}
    tree.postfix = infixpostfix(infix)
    stack := nodestack.Stack[*Node]{}
    str := strings.Split(tree.postfix, "")
    for index := 0; index < len(str); index++ {
        if str[index] >= string('a') && str[index] <=
                    string('z') {
            node := &Node{str[index], nil, nil}
            stack.Push(node)
        } else if (str[index] == "+") ||
                (str[index] == "-") ||
                (str[index] == "*") ||
```

```
                    (str[index] == "/") {
            right := stack.Top()
            stack.Pop()
            left := stack.Top()
            stack.Pop()
            node := &Node{str[index], nil, nil}
            node.left = left
            node.right = right
            stack.Push(node)
        }
    }
    tree.root = stack.Top()
    return tree
}

func (tree *ExpressionTree) Evaluate(node *Node,
        operandValues map[string]float64) float64 {
    if node == nil {
        return 0.0
    }
    if node.left == nil && node.right == nil {
        value := operandValues[node.ch]
        return value
    }
    leftValue := tree.Evaluate(node.left, operandValues)
    rightValue := tree.Evaluate(node.right, operandValues)
    if node.ch == "+" {
        return leftValue + rightValue
    } else if node.ch == "-" {
        return leftValue - rightValue
    } else if node.ch == "*" {
        return leftValue * rightValue
    } else {
        return leftValue / rightValue
    }
}
```

```go
// From Listing 5.7
func infixpostfix(infix string) (postfix string) {
    operators := []string{"+", "-", "*", "/",, ")"}
    postfix = ""
    nodeStack := nodestack.Stack[string]{}
    for index := 0; index < len(infix); index++ {
        newSymbol := string(infix[index])
        if newSymbol == " " || newSymbol == "\n" {
            continue
        }
        if newSymbol >= "a" && newSymbol <= "z" {
            postfix += newSymbol
        }
        if isPresent(newSymbol, operators) {
            if !nodeStack.IsEmpty() {
                topSymbol := nodeStack.Top()
                if precedence(topSymbol, newSymbol) ==
                        true {
                    if topSymbol != "(" {
                        postfix += topSymbol
                    }
                    nodeStack.Pop()
                }
            }
            if newSymbol != ")" {
                nodeStack.Push(newSymbol)
            } else {
                for {
                    if nodeStack.IsEmpty() == true {
                        break
                    }
                    ch := nodeStack.Top()
                    if ch != "(" {
                        postfix += ch
                        nodeStack.Pop()
```

```go
            } else {
                nodeStack.Pop()
                break
            }
        }
    }
}
    for {
        if nodeStack.IsEmpty() == true {
            break
        }
        if nodeStack.Top() != "(" {
            postfix += nodeStack.Top()
            nodeStack.Pop()
        }
    }
    return postfix
}

// From Listing 5.7
func precedence(symbol1, symbol2 string) bool {
    if (symbol1 == "+" || symbol1 == "-") &&
            (symbol2 == "(" || symbol2 == "/") {
        return false
    } else if (symbol1 == "(" && symbol2 != ")") ||
            symbol2 == "(" {
        return false
    } else {
        return true
    }
}

// From Listing 5.7
func isPresent(symbol string,operators []string) bool {
    for i := 0; i < len(operators); i++ {
```

```
        if symbol == string(operators[i]) {
            return true
        }
    }
    return false
}

func main() {
    operandValues := map[string]float64{"a": 5.0, "b":
                2.0, "c": 3.0, "d": 2.0,
                "f": 4.0, "g": 8, "h": 17, "y": 20,
                "x": 14, "z": 3}
    infix := "((a+b)+(- d)/(f+g)+ h))+ y / (x - z)"
    expressionTree := NewTree(infix)
    fmt.Println("Expression tree evaluates to: ",
                expressionTree.Evaluate(expressionTree.root,
                                        operandValues))
}
/* Output
Expression tree evaluates to:  25.90151515151515
*/
```

In the next section, we implement the *ShowTreeGraph* function for an expression tree.

13.3 Implementation of ShowTreeGraph

If we use the code from Chapter 8 for graphing a binary tree and apply it, as is, to an expression tree, we get the graph shown in Figure 13-2 for the tree produced in Listing 13-1.

Why the failure? An expression tree is a binary tree, so one would expect the code of Chapter 8 to work here.

The suite of code for graphically displaying a binary tree assumes that each node has a unique value field.

An expression tree fails this requirement because there are nodes with identical values. For example, how many nodes contain a "+" for their value? Many!

To fix the problem so that we can deploy the code to graph an expression tree, we concatenate a unique numerical tag, as a string, to each node's *ch* field. Then when we create labels; we extract only the first character from node.ch. In this way, we have forced each node to have a unique string representation while we build the tree.

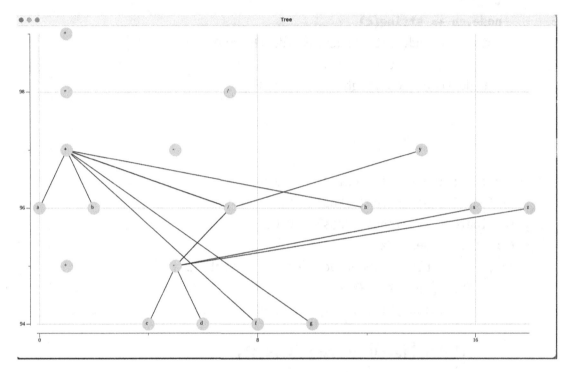

Figure 13-2. *Expression tree resulting from code in Chapter 8*

Listing 13-2 presents the revised portion of the suite of functions for graphing an expression tree. The four lines of code that are added are shown in boldface.

A variable *c* is defined global to function ***inorderLevel***. Each time this function is invoked, *c* is incremented by one, and ***node.ch*** is modified with this additional unique tag.

When adding labels in function ***ShowTreeGraph***, only the first character of ***node.ch*** is used, blocking out the unique tag.

Listing 13-2. Code for graphing an expression tree

```
var c  = 0

func inorderLevel(node *Node, level int) {
    if node != nil {
        inorderLevel(node.left, level + 1)
        c += 1
        node.ch += string(c)
        data = append(data, NodePos{node.ch, 100 -
                        level, -1})
        inorderLevel(node.right, level + 1)
    }
}

// Add Labels
for index := 0; index < len(data); index++ {
    x := float64(data[index].XPos) - 0.1
    y := float64(data[index].YPos) - 0.02
    str := data[index].Val
    label, err := plotter.NewLabels(plotter.XYLabels {
            XYs: []plotter.XY {
                {X: x ,Y: y},
            },
            Labels: []string{string(str[0])},
            },)
    if err != nil {
        log.Fatalf("could not creates labels
                    plotter: %+v", err)
    }
    p.Add(label)
}
```

When the modified suite of tree graphing functions is added to the code in Listing 13-2, the tree graph produced is shown in Figure 13-3.

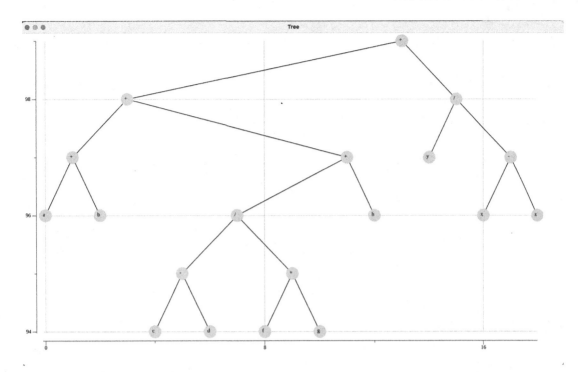

Figure 13-3. *Expression tree from modified code for graphing*

13.4 Summary

In this chapter, we implemented and discussed the details of building and evaluating an expression tree. We showed the modification needed for graphing an expression tree.

In the next chapter, we present a larger application that features concurrency.

## CHAPTER 14

Ecological Simulation with Concurrency

The previous chapter introduced expression trees. We showed how we can represent and evaluate simple mathematical expressions using such trees.

In this chapter, we switch gears. We present a concurrent implementation of an ecological simulation.

In the next section, we present an overview of the simulation.

14.1 Overview

This chapter presents an interesting emergent computation using a predator/prey model of a simple ecological system that simulates population dynamics. The design uses concurrency.

Many important concepts and techniques from previous chapters are used in this example. These include a graphical framework, extensive use of goroutines, object-oriented programming, type assertions (introduced in this chapter), implementing interfaces, and protecting shared data, to name a few.

We simulate the dynamics of three simplified marine-life species coexisting in an ocean with positions at any instant defined in a 50 × 50 grid of locations. At any moment, each of the 2500 locations contains nothing or a **shark** or a **tuna** or a **mackerel**.

In this simple food chain, **shark** is the top of the chain because **shark** can eat **tuna**. **Tuna** is second in the food chain because **tuna** can eat **mackerel**. **Mackerel** are at the bottom of the food chain. They are strictly a prey (can be eaten by **tuna**). **Tuna** is both a predator (can eat **mackerel**) and a prey (can be eaten by **shark**).

Each of the three species can reproduce according to rules to be specified. The two species that are predators (**shark** and **tuna**) can die of starvation. **Tuna** can also die because they are eaten by **shark**. All three species can die of old age. Since **shark** cannot

© Richard Wiener 2022
R. Wiener, *Generic Data Structures and Algorithms in Go*, https://doi.org/10.1007/978-1-4842-8191-8_14

be eaten, their population declines because of starvation (failure to eat **tuna** within a specified interval of time) or old age. Tuna also can reproduce and can die of starvation or old age. **Mackerel** can reproduce, die of old age, or die because they are eaten.

The rules of movement within the 50×50 grid of locations allow each critter (**shark**, **tuna,** or **mackerel**) to move concurrently once they are created. When they die (from starvation, old age, or being eaten), their movement stops, and they are purged from the ocean. Snapshots of the entire ocean are taken periodically to display the location of all the species along with empty locations.

We color-code each species, so the simulation output is most interesting as it shows the migration and population dynamics of the three species as a function of time. There is no communication between the critters. Each critter is an independent agent moving concurrently with all the other critters.

14.2 Specifications

We specify the rules that govern each of the three species.

Mackerel

A mackerel moves to an empty location in its immediate neighborhood (the collection of up to eight cells from the mackerel's current location, fewer if the mackerel is at one of the boundaries of the ocean (row 0, row 49, col 0, col 49)). If more than one empty location is found, it chooses one randomly and moves to this empty location, vacating its previous location. All mackerel, when created, are assigned a reproduction value. Each time it moves, its reproduction value is decremented by one. When its reproduction value becomes equal or less than zero, and the mackerel has been able to move to a neighboring empty location, it reproduces by creating a new mackerel and placing it in the location just vacated. This new mackerel takes on a life of its own and moves concurrently with the rest of the sea critters. If the mackerel was able to reproduce, its reproduction value is reset to its initial value. If the reproduction value is equal or less than zero but the mackerel was blocked from movement (no empty locations in its immediate neighborhood), it cannot reproduce on that move and must wait for a future move. Reproduction can occur only when the mackerel has moved, to allow the newly created mackerel to occupy the cell vacated by the mackerel that is reproducing.

If a mackerel is eaten by some tuna, it must be blocked from further moves because a dead mackerel cannot move or reproduce.

A mackerel is also assigned an age value when created. On each move, its age value is decremented by one. When the age value reaches 0, the mackerel dies. The dead mackerel must be blocked from further movement and purged from the ocean.

Tuna

The behavior of a tuna is only slightly different than a mackerel. On each move, its reproduction value, starvation value, and age value are decremented by one. If its starvation value or age value is zero, it dies and cannot move again and is purged from the ocean.

The tuna first attempts to move to a neighboring location containing a mackerel. If there is more than one mackerel found, it chooses one at random and moves to its location. The dead mackerel can no longer move and is purged from the ocean. The tuna's starvation value is reset to its original state. If there are no mackerel in the immediate neighborhood of the tuna, it attempts to move to a neighboring empty location, choosing a random empty cell if there is more than one. When its reproduction value is equal or less than 0, it reproduces using the same mechanism described for the mackerel. It cannot reproduce unless it has moved.

If some tuna is eaten by a shark, it cannot move again and must be purged from the ocean.

Shark

The behavior of a shark is like a tuna except that it cannot be eaten. When it moves, it first attempts to find and eat some tuna in one of its neighboring cells. Failing that, it moves to a neighboring empty location if one exists, choosing one randomly if more than one exists. Its rules for reproduction are identical to tuna and mackerel.

In summary, the population of mackerel increases because of reproduction. Its population decreases because of being eaten or old age.

The population of tuna increases because of reproduction. Its population decreases because of being eaten, starvation, or old age.

The population of shark increases because of reproduction. Its population decreases because of starvation or old age.

Output

Each critter is represented by a colored rectangle in the 50 × 50 grid of cells. Red rectangles represent shark. Blue rectangles represent tuna, and green rectangles represent mackerel. Empty cells are colored gray.

A census is conducted periodically, and the current positions of each critter and empty cells in the 50 × 50 grid are displayed graphically. This enables the migration pattern of each species to be dramatically displayed as the critters move concurrently.

A screenshot of the simulation in action is shown in Figure 14-1.

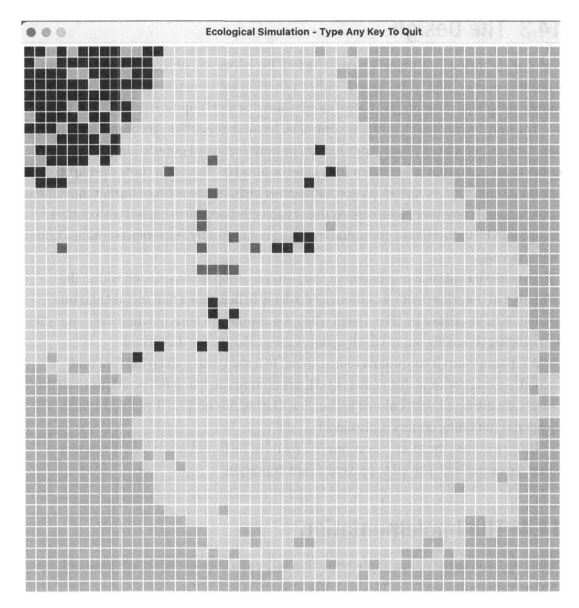

Figure 14-1. *Simulation in action*

Here, the population of mackerel has exploded outward, the population of tuna is about to encroach on the mackerel, and the sharks are waiting for the tuna to increase so they can feed on the tuna.

14.3 The Design

A global grid of location objects is constructed. Each location object contains an x and y position and a critter. This critter is either a shark, tuna, or mackerel. As each critter moves, the global location grid (a two-dimensional array) is updated.

The movement of each critter is controlled by an independent goroutine spawned when the critter is born (from reproduction or from the initial population). When the critter dies either from being eaten (mackerel or tuna) or starvation (tuna or shark) or old age (mackerel, tuna, or shark), its goroutine must be halted to prevent further movement and to control computer resources. As the ocean cells become occupied with critters, there could be thousands of goroutines running concurrently, each representing a critter that is moving.

To achieve continual movement of each critter, a loop is constructed within the goroutine of the critter, with a random sleep delay of between a half second and one second. This loop must be terminated when the simulation ends or when the critter dies. Terminating (breaking out of) the loop ends the goroutine for that critter.

A separate output goroutine is constructed in a loop with a sleep delay of one second. So every second, the census of critters is computed, and the positions of all the critters are displayed with colored rectangles. During this output, the global matrix of locations containing critters is displayed.

Using a mutex, the global locations matrix must be frozen when a critter moves or when the ocean is displayed to prevent a race condition.

14.4 The Implementation

Before presenting the entire implementation (over 400 lines of code), we show and discuss various pieces.

Data Model for Each Species

We start by examining the data model of each species and most importantly the global location matrix.

```
type Location struct {
    x       int
    y       int
```

```
    critter MarineLife
}

type MarineLife interface {
    Move()
    Reproduce(l Location)
    Starve() bool
    LifeOver() bool
}

type Tuna struct {
    repro int
    starv int
    life  int
    x, y int  // Set x to -1, y = -1 if dead
}

type Shark struct {
    repro int
    starv int
    life  int
    x, y int  // Set x to -1, y = -1 if dead
}

type Mackerel struct {
    repro int
    starv int
    life  int
    x, y int  // Set x to -1, y = -1 if dead
}

var locations [numRows][numCols]Location
```

Discussion of Code

Type *location* specifies *critter* as type *MarineLife*. For this to work, each of the concrete critter types (shark, tuna, and mackerel) must implement the *MarineLife* interface. This means that each of the concrete types must implement methods **Move**, **Reproduce**, **Starve**, and **LifeOver**.

Each of the critter types is defined by a struct containing the fields *repro*, *starv*, *life*, *x*, and *y*.

The global *locations* two-dimensional array is defined as containing *Location* objects.

Support Functions

Several support functions are defined that are needed to implement the *MarineLife* interface methods. These are shown as follows:

```go
func init() {
    rand.Seed(time.Now().UTC().UnixNano())
}

func distanceOfOne(x1, y1, x2, y2 float64) bool {
    return (math.Abs(x2-x1) == 0 &&
            math.Abs(y2-y1) == 1) ||
        (math.Abs(x2-x1) == 1 && math.Abs(y2-y1) == 0)
        || (math.Abs(x2-x1) == 1 &&
            math.Abs(y2-y1) == 1)
}

func initializeLocations() {
    for row := 0; row < numRows; row++ {
        for col := 0; col < numCols; col++ {
            locations[row][col] =
                Location{col, row, nil}
        }
    }
}

func findRandomCritter(x int, y int,
        critter MarineLife) (bool, Location) {
    // Send in nil for critter to get random empty
    // location
    result := []Location{}
    for r := 0; r < numRows; r++ {
```

```
    for c := 0; c < numCols; c++ {
        d := distanceOfOne(float64(x), float64(y),
            float64(c), float64(r))
        if d == true &&
        reflect.TypeOf(locations[c][r].critter) ==
        reflect.TypeOf(critter) {
            result = append(result, Location{r, c,
                    critter})
        }
    }
}
if len(result) == 0 {
    return false, Location{}
} else {
    return true, result[rand.Intn(len(result))]
}
}
```

Discussion of Code

We use the ***reflect.TypeOf*** method in function ***findRandomCritter*** to create a slice of ***Location*** objects containing the critter that is input to this function. This function returns two outputs and allows the caller to determine whether a target location has been found.

Function ***init()*** seeds the random number generator with the current clock time that assures different results each time the simulation is run.

Function ***initializeLocations*** assigns an *x* and *y* coordinate to each cell and assigns each cell with a ***nil*** critter.

Required Methods for Mackerel to Be of Type MarineLife

```
func (mackerel *Mackerel) Move() {
    for ; quit == false ; {
        if mackerel.x == -1 { // mackerel has been
                        // killed
```

```go
            break
        }
        mutex.Lock()
        mackerel.repro -= 1
        mackerel.starv -= 1
        mackerel.life -= 1
        if mackerel.LifeOver() || mackerel.Starve() {
            locations[mackerel.y][mackerel.x].critter
                        = nil
            mackerel.x = -1
            mackerel.y = -1
            mutex.Unlock()
            break
        }
        // Find random neighbor that has no critter
        found, newLoc := findRandomCritter(mackerel.x,
                mackerel.y, nil)
        if found == true {
            fmt.Printf("\nMackerel Move from <%d, %d>
                to <%d, %d>", mackerel.x,
                mackerel.y, newLoc.x, newLoc.y)
            mackerel.Reproduce(newLoc)
        }
        mutex.Unlock()
        time.Sleep(time.Duration(rand.Intn(500) + 500)
                        * time.Millisecond)
    }
}

func (mackerel Mackerel) Starve() bool {
    return mackerel.starv <= 0
}

func (mackerel Mackerel) LifeOver() bool {
    return mackerel.life <= 0
}
```

```
func (mackerel *Mackerel) Reproduce(l Location) {
    if mackerel.x == -1 {
        return
    }
    if mackerel.repro <= 0 {
        newMackerel := new(Mackerel)
        newMackerel.repro = MACKERELREPRO
        newMackerel.starv = MACKERELSTARVE
        newMackerel.life = MACKERELLIFE
        newMackerel.x = mackerel.x
        newMackerel.y = mackerel.y
        locations[mackerel.y][mackerel.x].critter =
                            newMackerel
        go newMackerel.Move()
    } else {
        locations[mackerel.y][mackerel.x].critter = nil
    }
    mackerel.x = l.x // assign mackerel to new location
    mackerel.y = l.y
    // add mackerel to new location
    locations[l.y][l.x].critter = mackerel
}
```

Discussion of Code

The **Move** method for mackerel takes a pointer to a **Mackerel** as receiver of the method. This is needed since the **mackerel** receiver may have its internal data modified.

In the for-loop that defines successive moves, if the x coordinate of the mackerel object is negative 1, we break out of the loop, which terminates the method. This method will be defined elsewhere as a goroutine.

We lock the mutex to prevent the global **locations** matrix from being changed outside of this goroutine. We decrement the three fields **repro**, **starv**, and **life**.

If either **Starve** or **LifeOver** is true, we purge the mackerel object from the ocean (setting its critter value to nil at the appropriate **location[mackerel.y][mackerel.x]**). We terminate the goroutine of the dead mackerel object by setting its x and y values to -1. We unlock the mutex.

If an empty target location is found, we output the move to the console and pass the *newLoc* to the *Reproduce* method. We unlock the mutex. We pause the goroutine loop using a random sleep interval.

The *Reproduce* method uses a pointer receiver since the receiver's internal data may be changed.

If the *repro* value is equal or less than zero, we create a new mackerel object using global constants that define the initial field values *repro*, *starv*, and *life*. We assign the new mackerel object to the **critter** field of **Location** and its x and y values to the location vacated by the reproducing mackerel.

If the *repro* value is greater than one, we set the vacated location to a critter value of nil.

Finally, we set the *x* and *y* coordinates of the mackerel to the new location.

Move Method for Shark

We next show the implementation of the *Move* method for **Shark**.

```
func (shark *Shark) Move() {
    for ; quit == false ; {
        if shark.x == -1 { // Shark no longer alive
            break
        }
        mutex.Lock()
        shark.repro -= 1
        shark.starv -= 1
        shark.life -= 1
        if shark.LifeOver() || shark.Starve() {
            locations[shark.y][shark.x].critter = nil
            shark.x = -1
            shark.y = -1
            mutex.Unlock()
            break
        }
        // Find random neighbor that has tuna
        found, newLoc := findRandomCritter(shark.x,
                            shark.y, new(Tuna))
```

```
if found == true {
    fmt.Printf("\nShark Move from <%d, %d> to
            <%d, %d>", shark.x, shark.y, newLoc.x,
            newLoc.y)
    shark.starv = SHARKSTARVE

    // Type assertion
    eatenTuna := locations[newLoc.y][newLoc.x].critter.(*Tuna)

    // Must stop go routine for tuna that was
    // eaten
    eatenTuna.x = -1
    eatenTuna.y = -1
    fmt.Printf("\nEaten tuna = %v", eatenTuna)
    shark.Reproduce(newLoc)
} else {
    found, newLoc = findRandomCritter(shark.x,
        shark.y, nil)
    if found == true {
        fmt.Printf("\nShark Move from <%d, %d>
            to <%d, %d>", shark.x,
        shark.y, newLoc.x, newLoc.y)
        shark.Reproduce(newLoc)
    }
}
mutex.Unlock()
time.Sleep(time.Duration(rand.Intn(500) + 500)
        * time.Millisecond)
    }
}
```

Discussion of Code

Most of the implementation details of **Move** for **Shark** are the same as for **Mackerel**. The only change is that the shark first looks for a neighboring tuna to eat.

Here, we encounter a **type assertion**. Let's look closely at this.

We invoke the *findRandomCritter* method as follows passing *new(Tuna)* as the third parameter:

```
found, newLoc := findRandomCritter(shark.x, shark.y,
                new(Tuna))
```

If found is true, we set the *starv* value back to its *SHARKSTARVE* initial value. Then we assign the variable *eatenTuna* as follows:

```
// Type assertion
eatenTuna := locations[newLoc.y][newLoc.x].critter.(*Tuna)
```

This type assertion asserts that **locations[[newLoc.y][newLoc.x]** is of type ***Tuna**.

Since this assertion is true, we can treat *eatenTuna* as if it had been defined to be of type ***Tuna**.

By setting the *x* and *y* values of *eatenTuna* to -1, we effectively terminate the goroutine for the eaten **Tuna** object.

Type assertions of this kind are useful when it is necessary to act on the actual type of an object whose formal type is an interface.

It is essential that the **Tuna** type implement the *MarineLife* interface for this to work. It does!

The other three methods that implement the *MarineLife* interface for type **Shark** are the same.

Move Method for Tuna

The *Move* method for class **Tuna** is essentially the same as the *Move* method just described for type **Shark**.

```
func (tuna *Tuna) Move() {
    for ; quit == false ; {
        if tuna.x == -1 { // Tuna no longer alive
            break
        }
        mutex.Lock()
        tuna.repro -= 1
        tuna.starv -= 1
        tuna.life -= 1
```

```go
if tuna.LifeOver() || tuna.Starve() {
    locations[tuna.y][tuna.x].critter = nil
    tuna.x = -1
    tuna.y = -1
    mutex.Unlock()
    break
}
// Find random neighbor that is a Mackerel
found, newLoc := findRandomCritter(tuna.x,
                tuna.y, new(Mackerel))
if found == true {
    fmt.Printf("\nTuna Move from <%d, %d> to
        <%d, %d>", tuna.x, tuna.y, newLoc.x,
        newLoc.y)
    tuna.starv = TUNASTARVE
    // Must stop go routine for mackerel that
    // was eaten
    // Type assertion
    eatenMackerel:=         locations[newLoc.y][newLoc.x].critter.
                            (*Mackerel)
    eatenMackerel.x = -1
    eatenMackerel.y = -1
    fmt.Printf("\nEaten mackerel = %v",
        eatenMackerel)
    tuna.Reproduce(newLoc)
}
found, newLoc = findRandomCritter(tuna.x,
                tuna.y, nil)
if found == true {
    fmt.Printf("\nTuna Move from <%d, %d> to
        <%d, %d>", tuna.x, tuna.y, newLoc.x,
        newLoc.y)
    tuna.Reproduce(newLoc)
}
mutex.Unlock()
```

```
        time.Sleep(time.Duration(rand.Intn(500) + 500)
                    * time.Millisecond)
    }
}
```

A similar **type assertion** is used to enable the killing of the eaten mackerel.

Output Function for the Graphical Display of Critters

The *output* function that produces a graphical display of the critters is given as follows:

```
func output() *fyne.Container {
    for col := 0; col < numCols; col++ {
        for row := 0; row < numRows; row++ {
            if locations[col][row].critter == nil {
                rect =
                    canvas.NewRectangle(&color.RGBA{B:
                        200, R: 200, G: 200, A: 255})
            } else if
        reflect.TypeOf(locations[col][row].critter) ==
        reflect.TypeOf(new(Tuna)) {
                rect =
                    canvas.NewRectangle(&color.RGBA{B:
                        255, R: 0, G: 0, A: 255})
            } else if
        reflect.TypeOf(locations[col][row].critter) ==
        reflect.TypeOf(new(Shark)) {
                rect =
                    canvas.NewRectangle(&color.RGBA{B:
                        0, R: 255, G: 0, A: 255})
            } else if
        reflect.TypeOf(locations[col][row].critter) ==
        reflect.TypeOf(new(Mackerel)) {
                rect =
                    canvas.NewRectangle(&color.RGBA{B:
```

```
                    0, R: 0, G: 255, A: 255})
        }
        rect.Resize(fyne.NewSize(10, 10))
        rect.Move(fyne.NewPos(float32(col * 11),
                float32(row * 11)))
        segments[col + numCols * row] = rect
    }
}
return container.NewWithoutLayout(segments...)
}
```

It is supported by the following global declarations:

```
const (
    numRows int = 50
    numCols int = 50
    MAKERELREPRO int = 4
    MAKERELSTARVE int = 10000000
    MAKERELLIFE int = 30
    TUNAREPRO int = 8
    TUNASTARVE int = 11
    TUNALIFE int = 18
    SHARKREPRO int = 15
    SHARKSTARVE int = 25
    SHARKLIFE int = 30
)

var (
    quit        bool
    contain     *fyne.Container
    rect        *canvas.Rectangle
    mutex = &sync.Mutex{}
    // Holds rectangle objects
    segments  = make([]fyne.CanvasObject, numRows *
                    numCols)
)
```

The *output* function is contained within the following goroutine in function **main**:

```go
go func() {
    for ; ; {
        mutex.Lock()
        contain := output()
        mutex.Unlock()
        w.SetContent(contain)
        time.Sleep(1000 * time.Millisecond)
    }
}()
```

Discussion of Code

In a loop that queries every location object, a rectangle, *rect*, is defined with its color based on the type of critter occupying the location. These rectangles are assigned to the segments array that allows *w.SetContent* to display the rectangles.

Full Implementation of Simulation

The implementation of the ecological simulation is presented in Listing 14-1. Functions presented and discussed previously are snipped out in the interest of space. You can download the full source code from the website specified in the Preface and run the simulation.

Listing 14-1. Ecological simulation

```go
package main

import (
    "fmt"
    "math"
    "math/rand"
    "reflect"
    "time"
    "image/color"
    "fyne.io/fyne/v2"
```

```go
    "fyne.io/fyne/v2/app"
    "fyne.io/fyne/v2/canvas"
    "fyne.io/fyne/v2/container"
    "sync"
)

const (
    numRows int = 50
    numCols int = 50
    MAKERELREPRO int = 4
    MAKERELSTARVE int = 10000000
    MAKERELLIFE int = 30
    TUNAREPRO int = 8
    TUNASTARVE int = 11
    TUNALIFE int = 18
    SHARKREPRO int = 15
    SHARKSTARVE int = 25
    SHARKLIFE int = 30
)

var (
    quit        bool
    contain     *fyne.Container
    rect        *canvas.Rectangle
    mutex =     &sync.Mutex{}
    // Holds rectangle objects
    segments  = make([]fyne.CanvasObject, numRows *
                numCols)
)

type Location struct {
    x       int
    y       int
    critter MarineLife
}

type MarineLife interface {
```

```go
    Move()
    Reproduce(l Location)
    Starve() bool
    LifeOver() bool
}

type Tuna struct {
    repro int // Moves til reproduction
    starv int // Movew til starvation
    life  int // Moves til life over
    x, y int  // Set x to -1, y = -1 if dead
}

type Shark struct {
    repro int
    starv int
    life  int
    x, y int  // Set x to -1, y = -1 if dead
}

type Mackerel struct {
    repro int
    starv int
    life  int
    x, y int  // Set x to -1, y = -1 if dead
}

var locations [numRows][numCols]Location

func init() {
    rand.Seed(time.Now().UTC().UnixNano())
}

func distanceOfOne(x1, y1, x2, y2 float64) bool {
    // snip
}
```

```go
func initializeLocations() {
    // snip
}

func findRandomCritter(x int, y int, critter MarineLife) (bool, Location) {
    // snip
}

func (tuna *Tuna) Move() {
    // snip
}

func (shark *Shark) Move() {
    // snip
}

func (mackerel *Mackerel) Move() {
    // snip
}

func (tuna Tuna) Starve() bool {
    // snip
}

func (tuna Tuna) LifeOver() bool {
    // snip
}

func (shark Shark) Starve() bool {
    // snip
}

func (shark Shark) LifeOver() bool {
    // snip
}

func (mackerel Mackerel) Starve() bool {
    // snip
}
```

```go
func (mackerel Mackerel) LifeOver() bool {
    // snip
}

func (tuna *Tuna) Reproduce(l Location) {
    // snip
}

func (shark *Shark) Reproduce(l Location) {
    // snip
}

func (mackerel *Mackerel) Reproduce(l Location) {
    // snip
}

func output() *fyne.Container {
    // snip
}

func main() {
    quit = false
    a := app.New()
    w := a.NewWindow("Ecological Simulation - Type Any
            Key To Quit")
    w.Resize(fyne.NewSize(600, 600))
    w.SetFixedSize(true)

    initializeLocations()

    newTuna := new(Tuna)
    newTuna.repro = TUNAREPRO
    newTuna.starv = TUNASTARVE
    newTuna.life = TUNALIFE
    newTuna.x = 15
    newTuna.y = 15
    locations[15][15].critter = newTuna
    go newTuna.Move()
```

```
newTuna = new(Tuna)
newTuna.repro = TUNAREPRO
newTuna.starv = TUNASTARVE
newTuna.life = TUNALIFE
newTuna.x = 19
newTuna.y = 19
locations[19][19].critter = newTuna
go newTuna.Move()

newTuna = new(Tuna)
newTuna.repro = TUNAREPRO
newTuna.starv = TUNASTARVE
newTuna.life = TUNALIFE
newTuna.x = 4
newTuna.y = 4
locations[4][4].critter = newTuna
go newTuna.Move()

newShark := new(Shark)
newShark.repro = SHARKREPRO
newShark.starv = SHARKSTARVE
newShark.life = SHARKLIFE
newShark.x = 11
newShark.y = 11
locations[11][11].critter = newShark
go newShark.Move()

newShark = new(Shark)
newShark.repro = SHARKREPRO
newShark.starv = SHARKSTARVE
newShark.life = SHARKLIFE
newShark.x = 16
newShark.y = 16
locations[16][16].critter = newShark
go newShark.Move()

newMackerel := new(Mackerel)
```

```
newMackerel.repro = MAKERELREPRO
newMackerel.starv = MAKERELSTARVE
newMackerel.life = MAKERELLIFE
newMackerel.x = 2
newMackerel.y = 2
locations[2][2].critter = newMackerel
go newMackerel.Move()

newMackerel = new(Mackerel)
newMackerel.repro = MAKERELREPRO
newMackerel.starv = MAKERELSTARVE
newMackerel.life = MAKERELLIFE
newMackerel.x = 13
newMackerel.y = 8
locations[8][13].critter = newMackerel
go newMackerel.Move()

newMackerel = new(Mackerel)
newMackerel.repro = MAKERELREPRO
newMackerel.starv = MAKERELSTARVE
newMackerel.life = MAKERELLIFE
newMackerel.x = 16
newMackerel.y = 16
locations[16][16].critter = newMackerel
go newMackerel.Move()

newMackerel = new(Mackerel)
newMackerel.repro = MAKERELREPRO
newMackerel.starv = MAKERELSTARVE
newMackerel.life = MAKERELLIFE
newMackerel.x = 28
newMackerel.y = 28
locations[28][28].critter = newMackerel
go newMackerel.Move()

go func() {
    for ; ; {
```

```
            mutex.Lock()
            contain := output()
            mutex.Unlock()
            w.SetContent(contain)
            time.Sleep(1000  * time.Millisecond)
        }
    }()

    w.Canvas().SetOnTypedKey(func(k *fyne.KeyEvent) {      // Shuts down
                                                             simulation

        quit = true
        w.Close()
    })

    w.ShowAndRun()
}
```

14.5 Summary

A concurrent implementation of an ecological simulation is presented in this chapter. Type assertions are introduced and used in the implementation.

Many important concepts and techniques from previous chapters are used in this example. These include a graphical framework, extensive use of goroutines, object-oriented programming, type assertions, implementing interfaces, and protecting shared data.

In the next chapter, we introduce an important technique of algorithm design, dynamic programming.

CHAPTER 15

Dynamic Programming

The previous chapter presented a concurrent implementation of an ecological simulation. It used many of the techniques presented earlier in this book.

This chapter changes focus from data structures to algorithm design.

We introduce an algorithmic technique for solving optimization problems, dynamic programming, and apply this technique to several problems.

As you will see in this chapter, "if you cannot remember the past, you are destined to repeat it."

In the next section, we present a simple example of dynamic programming, the computation of the nth Fibonacci number. We explore two dynamic programming approaches.

15.1 Example of Dynamic Programming: nth Fibonacci Number

The central mechanism of dynamic programming is representing the solution to a problem in terms of smaller subproblems, each of which has optimal solutions. Each subproblem is a smaller version of the original problem. By storing the results to the smaller problems, we can efficiently obtain the results to the larger problem.

A simple example involves the computation of the nth Fibonacci number.

$$\text{Fib}(n) = \text{Fib}(n\text{-}1) + \text{Fib}(n\text{-}2), \text{ for } n > 1$$

The first two numbers in the sequence are 0 and 1.
The initial sequence of Fibonacci numbers is

$$[0, 1, 1, 2, 3, 5, 8, 13, 21, ...]$$

We examine three alternative algorithms for computing the nth Fibonacci number. The first two involve dynamic programming, and the third involves recursion.

427

© Richard Wiener 2022
R. Wiener, *Generic Data Structures and Algorithms in Go*, https://doi.org/10.1007/978-1-4842-8191-8_15

Top-Down Dynamic Programming

Consider the function *FibonacciTopDown* and its support function *computeFromCache* given as follows:

```go
func FibonacciTopDown(n int) int64 {
    firstTwoCases := map[int]int64{
        0: 0,
        1: 1,
    }
    return computeFromCache(n, firstTwoCases)
}

func computeFromCache(n int, cache map[int]int64) int64 {
    // If answer already found for n, return it
    if val, found := cache[n]; found {
        return val
    }
    cache[n] = computeFromCache(n - 1, cache) +
               computeFromCache(n - 2, cache)
    return cache[n]
}
```

A **map** is used in *computeFromCache* to return a solution if it has already been calculated.

The variable *cache* holds the key-value pairs (n and the nth Fibonacci number).

This is dynamic programming because a problem of size n is computed in terms of problems of size n – 1 and n – 2.

The computational complexity of this top-down approach is **O(n)**. The space complexity is also O(n) because of the map that holds previous computations.

Bottom-Up Dynamic Programming

Consider the function *FibonacciBottomUp* presented as follows:

```go
func FibonacciBottomUp(n int) int64 {
    table := []int64{0, 1}
    for i := 2; i <= n; i++ {
```

```
        table = append(table, table[i - 1] +
                    table[i - 2])
    }
    return table[n]
}
```

We construct the variable *table* from **0** to **n**, bottom-up.

The computational complexity of this solution is also **O(n)**. The space complexity is **O(1)**.

Recursive Solution

The function *Fib*, presented in the following, is a recursive solution. But it is of computational complexity **O(2^n)**. This is intractable.

```
func Fib(n int64) int64 {
    if n < 2 {
        return n
    }
    return Fib(n - 1) + Fib(n - 2)
}
```

Listing 15-1 presents the three approaches along with a main driver that does a timing analysis.

Listing 15-1. Fibonacci numbers

```
package main

import (
    "fmt"
    "time"
)

func FibonacciTopDown(n int) int64 {
    firstTwoCases := map[int]int64{
        0: 0,
        1: 1,
    }
```

```go
    return computeFromCache(n, firstTwoCases)
}

func computeFromCache(n int, cache map[int]int64) int64 {
    // If answer already found for n, return it
    if val, found := cache[n]; found {
        return val
    }
    cache[n] = computeFromCache(n - 1, cache) +
                computeFromCache(n - 2, cache)
    return cache[n]
}

func FibonacciBottomUp(n int) int64 {
    table := []int64{0, 1}
    for i := 2; i <= n; i++ {
        table = append(table, table[i - 1] +
                    table[i - 2])
    }
    return table[n]
}

func Fib(n int64) int64 {
    if n < 2 {
        return n
    }
    return Fib(n - 1) + Fib(n - 2)
}

func main() {
    fmt.Println("fib(7) = ", FibonacciTopDown(7))
    start := time.Now()
    fib40 := FibonacciTopDown(40)
    elapsed := time.Since(start)
    fmt.Println("Value of FibonacciTopDown(40): ", fib40)
    fmt.Println("Computation time: ", elapsed)

    fmt.Println("fib(7) = ", FibonacciBottomUp(7))
```

```
    start = time.Now()
    fib40 = FibonacciBottomUp(40)
    elapsed = time.Since(start)
    fmt.Println("\nValue of FibonacciBottomUp(40): ", fib40)
    fmt.Println("Computation time: ", elapsed)

    fmt.Println("fib(7) = ", Fib(7))
    start = time.Now()
    fib40 = Fib(40)
    elapsed = time.Since(start)
    fmt.Println("\nValue of Fib(40): ", fib40)
    fmt.Println("Computation time: ", elapsed)
}
/* Output
fib(7) =   13
Value of FibonacciTopDown(40):   102334155
Computation time:   36.136µs
fib(7) =   13

Value of FibonacciBottomUp(40):   102334155
Computation time:   7.377µs
fib(7) =   13
Value of Fib(40):   102334155
Computation time:   424.44211ms
*/
```

Discussion of Code

The dynamic programming bottom-up approach is roughly five times faster than the dynamic programming top-down approach. Both are significantly faster than the recursive approach.

In the next section, we examine a classic problem from algorithm design, the 0/1 knapsack problem.

15.2 Another Application: 0/1 Knapsack Problem

Suppose we are given a set of objects. We wish to pack a subset of these objects into a knapsack with a specified weight limit. Each object to be considered has a specified weight and profit, if included in the knapsack. We wish to choose a subset of the objects that maximizes our profit.

As a small example, suppose the four potential objects have weights **4, 6, 2, 8** and profits **12, 15, 9, 21**. Suppose the weight limit on the knapsack is 10.

Let us enumerate combinations of objects whose total weight is <= 10.

Object1 + Object2 (total weight 10), profit 27

Object1 + Object3 (total weight 6), profit 21

Object2 + Object3 (total weight 8), profit 24

Object3 + Object4 (total weight 10), profit 30

The optimum solution is to include Object3 and Object4 in the knapsack.

Brute-Force Solution

A brute-force solution enumerates every combination of subsets of weights and profits. Consider the following function:

```
// Brute Force solution
func KnapSackBF(weightLimit int, weights []int, profits []int, n int) int {
    if n == 0 || weightLimit == 0 {
        return 0
    }
    if weights[n - 1] > weightLimit {
        return KnapSackBF(weightLimit, weights, profits, n - 1)
    } else {
        // Assume that we include object n - 1
        value1 := profits[n - 1] +
                    KnapSackBF(weightLimit -
            weights[n - 1], weights, profits, n - 1)
        // Assume that we do not include object n - 1
        value2 := KnapSackBF(weightLimit, weights,
                    profits, n - 1)
        if value1 >= value2 {
```

```
            return value1
        } else {
            return value2
        }
    }
}
```

Discussion of Code

If the weight at **weights[n – 1]** exceeds *weightLimit*, recursively invoke *KnapSackBF* replacing **n** by **n – 1**.

Otherwise, compute *value1*, which assumes that you include object **n – 1**, and *value2*, which assumes that you exclude object **n – 1**. Return the larger of *value1* and *value2* at this level of recursion.

This brute-force algorithm is our first example of a computationally intractable procedure.

Since all the subsets are encompassed and used in this recursive function, and it is well known that the number of subsets of a set of size **n** is 2^n, we conclude that this brute-force method is $O(2^n)$. The computation time grows exponentially, asymptotically.

Dynamic Programming Solution

A dynamic programming solution to this problem is given as follows:

```
// Dynamic Programming solution
func KnapSackDP(weightLimit int, weights []int, profits
      []int) int {
   n := len(weights)
   if weightLimit <= 0 || n == 0 || len(profits) != n    {
       return 0
   }
   // Create a (n + 1 x weighlimit + 1) table
   table := make([][]int, n + 1)
   for row := 0; row < n + 1; row++ {
       table[row] = make([]int, weightLimit + 1)
   }
```

```
    for i := 0; i < n + 1; i++ {
        for w := 0; w < weightLimit + 1; w++ {
            if i == 0 || w == 0 {
                table[i][w] = 0
            } else if weights[i - 1] <= w {
                // Include item i
                wt := w - weights[i - 1]
                profit1 := profits[i - 1] +
                            table[i - 1][wt]
                // Exclude item i
                profit2 := table[i  - 1][w]
                if profit1 >= profit2 {
                    table[i][w] = profit1
                } else {
                    table[i][w] = profit2
                }
            } else {
                // Exclude item
                table[i][w] = table[i - 1][w]
            }
        }
    }
    return table[n][weightLimit]
}
```

Discussion of Code

We utilize a two-dimensional slice to accomplish the dynamic programming. Let us "walk" through a small example to see how the algorithm works.

Suppose our weights and profits arrays are as follows:

Weights = [3, 5, 1]

Profits = [10, 20, 1]

WeightLimit = 5

We compute **n** equal to 2.

We create a 4 × 6 table.

The following table is generated in **KnapsackDP**:

```
0 0 0   0   0   0
0 0 0   0  10  10
0 0 0  10  10  20
0 1 1  10  11  20
```

Because the dynamic programming solution is found using two nested loops, the computation is **O(n x L)** complexity where L is the weight limit.

In Listing 15-2, we compare the computation time for each of the algorithms for solving the 0/1 knapsack problem.

Listing 15-2. 0/1 Knapsack computation times

```go
package main

import (
    "fmt"
    "time"
)

// Brute Force solution - Snip


// Dynamic Programming solution - Snip

func main() {
    weights := []int{4, 6, 2, 8}
    profits := []int{12, 15, 9, 21}
    fmt.Println("Solution 1 = ", KnapSackBF(10, weights, profits, 4))

    weights1 := []int{4, 6, 2, 8, 1, 17, 23, 10, 4, 8}
    profits1 := []int{12, 15, 9, 21, 5, 8, 20, 6, 1, 15}
    result := KnapSackBF(20, weights1, profits1, 10)
    fmt.Println("Solution 2 = ", result)

    weights2 := []int{}
    for i := 0; i < 800; i++ {
        weights2 = append(weights2, 2 * i)
    }
    profits2 := []int{}
    for i := 0; i < 800; i++ {
```

```
        profits2 = append(profits2, 3 * i)
    }

    start := time.Now()
    result2 := KnapSackBF(400, weights2, profits2, 800)
    elapsed := time.Since(start)
    fmt.Println("Solution 3 = ", result2)
    fmt.Println("Time for solution3 (brute force): ",
                elapsed)

    start = time.Now()
    result3 := KnapSackDP(400, weights2, profits2)
    elapsed = time.Since(start)
    fmt.Println("Solution 3 = ", result3)
    fmt.Println("Time for solution3 (dynamic programming): ", elapsed)
}
/* Output
Solution 1 = 30
Solution 2 = 57
Solution 3 = 600
Time for solution3 (brute force): 1m10.248200934s
Solution 3 = 600
Time for solution3 (dynamic prograamming): 1.621038ms
*/
```

Discussion of Code

For a problem involving 800 weights and profits, the dynamic programming solution is 43,000 times faster than the brute-force solution. If the weight limit in this problem were increased, the computation time for the brute-force solution would become intractable.

In the next section, we apply dynamic programming to finding the longest subsequence in two DNA strings.

15.3 DNA Subsequences

DNA strings are a sequence of characters taken from the alphabet {A, C, G, T}. An example of such a string is "**CGT**TACAA**TTT**GC**G**".

We define a **subsequence** of a string to be a sequence of characters taken in order (not necessarily contiguous order) from the characters of the original string as one scans from left to right in the original string.

For the preceding string, a subsequence would be "GTAAAGG". This sequence is taken from the characters shown in boldface from the original string.

In computational genetics, an important problem is finding the longest subsequence that is common to two DNA strings. This is the longest subsequence problem.

A brute-force approach would be to enumerate all the subsequences of string1 and then test each one against string2. If string1 has **n** characters and string2 has **m** characters, it would take $O(2^n m)$ for this brute-force algorithm. This is computationally intractable for a large string1.

We use dynamic programming to solve this problem.

We define **Length(j, k)** as the length of a longest string that is a subsequence of **X[0:j]** and **Y[0:k]**.

There are two cases to consider. In the first, X_{j-1} is equal to y_{k-1}.

It follows then that Length(j, k) = 1 + Length(j-1, k-1).

If x_{j-1} is not equal to y_{k-1}, we cannot have a subsequence that includes x_{j-1} and y_{k-12}.

We then set Length(j, k) = max{Length(j-1, k), Length(j, k - 1)}.

Length(j, 0) is 0 for j = 0, 1, ..., n and

Length(0, k) is 0 for k = 0, 1, ..., m

These recurrence relations give rise to a dynamic programming solution.

We create an $(n + 1) \times (m + 1)$ two-dimensional slice, L. We initialize this list to 0's. We iteratively construct L until we get $L_{n,m}$.

Listing 15-3 presents a dynamic programming solution to the longest common subsequence problem along with a main driver program with two test cases.

Listing 15-3. Longest common subsequence

```go
package main

import (
    "fmt"
)
```

```go
func max(value1, value2 int) int {
    if value1 >= value2 {
        return value1
    } else {
        return value2
    }
}

func reverse(x []rune) []rune {
    result := []rune{}
    for index := len(x) - 1; index >= 0; index-- {
        result = append(result, x[index])
    }
    return result
}

func longestCommonSubsequenceTable(x, y []rune) (LCS [][]int) {
    // Return matrix so that LCS[j][k] is longest
    // common sequence for x[0:j] and y[0:k]
    n := len(x)
    m := len(y)
    // Initialize LCS table of size (n + 1 x m + 1)
    LCS = make([][]int, n + 1)
    for row := 0; row < n + 1; row++ {
        LCS[row] = make([]int, m + 1)
    }
    for row := 0; row < n; row++ {
        for col := 0; col < m; col++ {
            if x[row] == y[col] {
                LCS[row + 1][col + 1] = 1 +
                            LCS[row][col]
            } else {
                LCS[row + 1][col + 1] =
                    max(LCS[row][col + 1],
                        LCS[row + 1][col])
            }
        }
    }
```

```go
    }
    return LCS
}

func LongestCommonSequence(x, y []rune) string {
    table := longestCommonSubsequenceTable(x, y)
    result := []rune{}
    j, k := len(x), len(y)
    for {
        if table[j][k] == 0 {
            break
        }
        if x[j - 1] == y[k - 1] {
            result = append(result, x[j - 1])
            j -= 1
            k -= 1
        } else if table[j - 1][k] >= table[j][k - 1] {
            j -= 1
            k -= 1
        }
    }
    return string(reverse(result))
}

func main() {
    x := "CGTTACAATTTGCG"
    y := "TTTTAAACGTGCG"
    lcs := LongestCommonSequence([]rune(x), []rune(y))
    fmt.Println(lcs)

    x = "ATCGAATTCCGGTAGTCGT"
    y = "CGATAGTTCAGCCAG"
    lcs = LongestCommonSequence([]rune(x), []rune(y))
    fmt.Println(lcs)
}
```

```
/* Output
TTAATGCG
TAGC
*/
```

Discussion of Code

In order to be able to access individual characters of the x and y input strings, we need to convert each to a slice of **rune**.

The recurrence relationships described earlier form the basis for the details in function *longestCommonSubsequenceTable*.

In function *LongestCommonSequence*, we start at the right of the table and work leftward toward the beginning of the table. Therefore, we need to reverse the result and convert the reversed slice of rune to a string.

The computational complexity of this dynamic programming solution is **O(n x m)**.

15.4 Summary

The algorithm design technique of dynamic programming was introduced. It was applied to several problems including Fibonacci numbers, 0/1 knapsack, and DNA subsequences. In all three problems, dynamic programming provides an efficient solution.

In the next chapter, we turn our attention to graph structures and some classic algorithms that utilize graphs.

CHAPTER 16

Graph Structures

In the previous chapter, we presented dynamic programming and three applications.

In this chapter, we introduce graph structures and some applications. We show several examples of how to represent a graph, and we examine some basic algorithms associated with graph traversal.

In the next section, we examine how graphs can be represented.

16.1 Representing Graphs

Graph data structures provide one of the most useful and powerful frameworks for algorithm design. A graph (not to be confused with a pictorial representation of a mathematical function) can represent a huge number of systems from communication networks, transportation, electrical grid, online interactions, games, and pattern matching, to name a few.

A graph consists of a set of nodes and edges between them: **Graph = (N, E)**, where N is the collection of nodes and E the collection of edges.

In a **directed graph**, each edge has a specified direction.

Nodes are **adjacent** if there is an edge connecting them. Nodes that are adjacent to a given node are **neighbors**.

The **degree** of a node is the number of nodes incident to it.

A **path** in a graph is a subgraph (subset of N and E) where the edges connect a series of nodes in a sequence without visiting any node more than once.

A **weighted graph** has a weight associated with each edge. The length of a path is the sum of its edge weights.

Consider the weighted directed graph shown in Figure 16-1.

R. Wiener, *Generic Data Structures and Algorithms in Go*, https://doi.org/10.1007/978-1-4842-8191-8_16

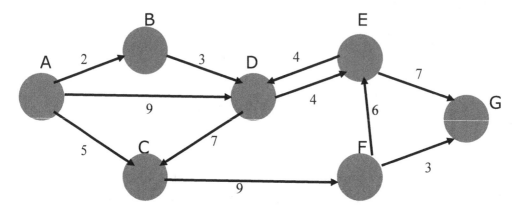

Figure 16-1. *A weighted graph*

In the next section, we discuss two methods for traversing this graph or any graph. We allow the node to be of generic type ***OrderedStringer***. Here, the generic type is ***String***.

16.2 Traversing Graphs

We look at two important traversal algorithms:

1. Depth-first search (DFS)

2. Breadth-first search (BFS)

The method **DFS** uses recursion and moves in a sequence outward from the starting node and continues to visit nodes in a sequence as far as possible from the starting node without revisiting a node and getting progressively closer to the starting node.

The method **BFS** uses iteration and an internal queue to traverse the graph incrementally from the starting node. Nodes are visited at distances progressively further from the starting vertex toward the furthest vertex from the starting vertex.

The edge values are stored in a global map variable as edges are defined.

In the next section, we discuss and implement a depth-first traversal and a breadth-first traversal of a graph with generic vertex values.

16.3 Depth- and Breadth-First Search

We start by defining important data types.

```
type OrderedStringer interface {
    comparable
    String() string
}

type Vertex[T OrderedStringer] struct {
    Key T
    Neighbors map[T]*Vertex[T]
}

type Graph[T OrderedStringer] struct {
    Vertices map[T]*Vertex[T]
}

var visitation []string
```

Our generic type is ***OrderedStringer***. Entities of this type must be comparable and have a string representation using the **String()** function.

A generic ***Vertex*** contains a ***Key*** (type **T**) and a map, ***Neighbors***, mapping keys (type T) to pointers to other vertices.

A generic ***Graph*** contains a map, ***Vertices***, mapping keys (type **T**) to pointers to vertices.

A global ***visitation*** variable is defined. This variable is a slice of string representing the traversal keys.

Several important functions and methods are shown as follows:

```
func NewVertex[T OrderedStringer](key T) *Vertex[T] {
    return &Vertex[T]{
        Key: key,
        Neighbors: map[T]*Vertex[T]{},
    }
}

func NewGraph[T OrderedStringer]() *Graph[T] {
    return &Graph[T]{Vertices: map[T]*Vertex[T]{}}
}
```

```go
func (g *Graph[T]) AddVertex(key T) {
    vertex := NewVertex(key)
    g.Vertices[key] = vertex
}

func (g *Graph[T]) AddEdge(key1, key2 T, edgeValue int) {
    vertex1 := g.Vertices[key1]
    vertex2 := g.Vertices[key2]
    if vertex1 == nil || vertex2 == nil {
        return
    }
    vertex1.Neighbors[vertex2.Key] = vertex2
    g.Vertices[vertex1.Key] = vertex1
    g.Vertices[vertex2.Key] = vertex2
}
```

The function *NewVertex* takes a key (of type **T**) and returns a pointer to a **Vertex** with an empty map, *Neighbors*.

The function *NewGraph* returns an empty graph with an empty map of *Vertices*.

The method *AddVertex* creates a new vertex and assigns *Vertices[key]* to the new vertex.

The method *AddEdge* takes the two keys, assigns these to the *Vertices* field of graph, and assigns the *Neighbors* field of *vertex1* to *vertex2*.

Depth-First Search

The first traversal method we examine is depth-first search. This method moves outward from the starting vertex and moves directly to the furthest vertex not yet visited.

It is implemented as follows:

```go
func (g *Graph[T]) DepthFirstSearch(start *Vertex[T],
        visited map[T]bool) {
    if start == nil {
        return
    }
    visited[start.Key] = true
    visitation = append(visitation, start.Key.String())
```

```
    // for each of the adjacent vertices, call the
    // function recursively if it hasn't yet been
    // visited
    for _, v := range start.Neighbors {
        // The sequence of v may change from run to run
        if visited[v.Key] {
            continue
        }
        g.DepthFirstSearch(v, visited)
    }
}
```

The parameter, *visited*, passed in must be initialized to an empty map before invoking the method.

The sequence of recursive calls causes the traversal to move away from the starting vertex until it is furthest away before backtracking and finding other vertices far away from the starting vertex.

Breadth-First Search

The second traversal method we examine is breadth-first search. This method visits vertices close to the starting vertex slowly moving outward and away from the starting vertex. There is no recursion used here. A queue is used to store vertices that neighbor visited vertices. These neighboring vertices are traversed first. So as the name of this method implies, this traversal moves to adjacent vertices slowly getting further from the starting vertex. It is implemented as follows:

```
type Queue[T any] struct {
    items []T
}

// Methods
func (queue *Queue[T]) Insert(item T) {
    // item is added to the right-most position in the
    // slice
    queue.items = append(queue.items, item)
}
```

```go
func (queue *Queue[T]) Remove() T {
    returnValue := queue.items[0]
    queue.items = queue.items[1:]
    return returnValue
}

func (g *Graph[T]) BreadthFirstSearch(start *Vertex[T],
            visited map[T]bool) {
    if start == nil {
        return
    }
    queue := Queue[*Vertex[T]]{} // Queue hold pointers
                                 // to Vertex
    current := start
    for {
        if !visited[current.Key] {
            visitation = append(visitation,
                        current.Key.String())
        }
        visited[current.Key] = true
        // Insert each neighboring vertex not visited
        // onto the queue
        for _, v := range current.Neighbors {
            if !visited[v.Key] {
                queue.Insert(v)
            }
        }
        // Grab first vertex in the queue and remove it
        if len(queue.items) > 0 {
            current = queue.Remove()
        } else {
            break
        }
    }
}
```

The *queue* plays a central role in the implementation of this method. It forces all nearby vertices to be visited early in contrast to depth-first search. Like the latter method, this method requires the parameter *visited* to be initialized to an empty map before invoking the method.

Listing 16-1 presents all the details of defining and traversing a graph. The graph shown in Section 16.1 is constructed in the main driver program.

Listing 16-1. Defining and traversing a graph

```go
package main

import (
    "fmt"
)

type OrderedStringer interface {
    comparable
    String() string
}

type Graph[T OrderedStringer] struct {
    Vertices map[T]*Vertex[T]
}

type Vertex[T OrderedStringer] struct {
    Key T
    Neighbors map[T]*Vertex[T]
}

var visitation []string

func NewVertex[T OrderedStringer](key T) *Vertex[T] {
    return &Vertex[T]{
        Key: key,
        Neighbors: map[T]*Vertex[T]{},
    }
}
```

```go
func NewGraph[T OrderedStringer]() *Graph[T] {
    return &Graph[T]{Vertices: map[T]*Vertex[T]{}}
}

func (g *Graph[T]) AddVertex(key T) {
    vertex := NewVertex(key)
    g.Vertices[key] = vertex
}

func (g *Graph[T]) AddEdge(key1, key2 T,
        edgeValue int) {
    vertex1 := g.Vertices[key1]
    vertex2 := g.Vertices[key2]
    if vertex1 == nil || vertex2 == nil {
        return
    }
    vertex1.Neighbors[vertex2.Key] = vertex2
    g.Vertices[vertex1.Key] = vertex1
    g.Vertices[vertex2.Key] = vertex2
}

func (g *Graph[T]) DepthFirstSearch(start *Vertex[T],
        visited map[T]bool) {
    if start == nil {
        return
    }
    visited[start.Key] = true
    visitation = append(visitation, start.Key.String())

    for _, v := range start.Neighbors {
        // The sequence of v may change from run to run
        if visited[v.Key] {
            continue
        }
        g.DepthFirstSearch(v, visited)
    }
}
```

```go
type Queue[T any] struct {
    items []T
}

// Methods
func (queue *Queue[T]) Insert(item T) {
    queue.items = append(queue.items, item)
}

func (queue *Queue[T]) Remove() T {
    returnValue := queue.items[0]
    queue.items = queue.items[1:]
    return returnValue
}

func (g *Graph[T]) BreadthFirstSearch(start *Vertex[T],
            visited map[T]bool) {
    if start == nil {
        return
    }
    queue := Queue[*Vertex[T]]{}
    current := start
    for {
        if !visited[current.Key] {
            visitation = append(visitation,
                            current.Key.String())
        }
        visited[current.Key] = true
        for _, v := range current.Neighbors {
            if !visited[v.Key] {
                queue.Insert(v)
            }
        }
        // Grab first vertex in the queue and remove it
        if len(queue.items) > 0 {
            current = queue.Remove()
        } else {
```

```go
                break
            }
        }
    }

    func (g *Graph[T]) String() string {
        result := ""
        for i := 0; i < len(visitation); i++ {
            result += " " + visitation[i]
        }
        return result
    }

    // Make String implement Stringer
    type String string

    func (str String) String() string {
        return string(str)
    }

    func main() {
        g := NewGraph[String]()
        g.AddVertex("A")
        start := g.Vertices["A"]
        g.AddVertex("B")
        g.AddVertex("C")
        g.AddVertex("D")
        g.AddVertex("E")
        g.AddVertex("F")
        g.AddVertex("G")
        g.AddEdge("A", "B", 2)
        g.AddEdge("A", "C", 5)
        g.AddEdge("A", "D", 9)
        g.AddEdge("B", "D", 3)
        g.AddEdge("C", "F", 9)
        g.AddEdge("D", "E", 4)
        g.AddEdge("E", "D", 4)
```

```
    g.AddEdge("F", "E", 6)
    g.AddEdge("E", "G", 7)
    g.AddEdge("F", "G", 3)

    visited := make(map[String]bool)
    visitation = []string{}
    g.DepthFirstSearch(start, visited)
    fmt.Println("Depth First Search:", g.String())
    visited = make(map[String]bool)
    visitation = []string{}
    g.BreadthFirstSearch(start, visited)
    fmt.Println("Breadth First Search:", g.String())
}
/* Output
Depth First Search:  A B D E G C F
Breadth First Search:  A C D B F E G
*/
```

The two output traversal sequences confirm our assertions about the sequence of visited nodes for depth- vs. breadth-first search.

In the next section, we present and implement a solution to a classic graph problem. We show how to find the shortest path from a source node to every other node in a graph.

16.4 Single-Source Shortest Path in Graph

Given a graph and a source node in the graph. Find the shortest paths from the source node to all the nodes in the graph.

The celebrated Dijkstra algorithm is presented. It was conceived by computer scientist Edsger W. Dijkstra in 1956 and published three years later.

Consider the graph with source node "A" shown in Figure 16-2.

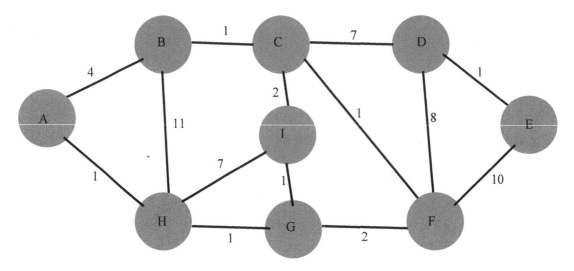

Figure 16-2. *Graph for single-source shortest path*

The Dijkstra algorithm returns the shortest distance from source node "A" to all the other nodes in the graph.

Implementation

Listing 16-2 presents an implementation of the Dijkstra algorithm.

Listing 16-2. Dijkstra algorithm

```
package main

import (
    "fmt"
    "github.com/jiaxwu/container/heap"
)

// PriorityQueue Priority queue
type PriorityQueue[T any] struct {
    h *heap.Heap[T]
}

func New[T any](less func(e1 T, e2 T) bool)
                *PriorityQueue[T] {
    return &PriorityQueue[T]{
```

```go
        h: heap.New(less),
    }
}

func (p *PriorityQueue[T]) Add(elem T) {
    p.h.Push(elem)
}

func (p *PriorityQueue[T]) Remove() T {
    return p.h.Pop()
}

func (p *PriorityQueue[T]) Len() int {
    return p.h.Len()
}

func (p *PriorityQueue[T]) Empty() bool {
    return p.Len() == 0
}

func Less(a, b tuple) bool {
    return a.weight < b.weight
}
// end priority queue

type edges = map[rune]int

type Graph map[rune]edges

type tuple struct {
    node    rune
    weight int
}

func convert(r rune) int {
    return int(r) - 65
}

func Dijkastra(graph Graph) []tuple {
```

```
distances := make([]tuple, len(graph))

for i := 0; i < len(graph); i++ {
    distances[i] = tuple{'A', 32767}
}

distances[0] = tuple{'A', 0}
heapQueue := New[tuple](Less)
t := tuple{'A', 0}
heapQueue.Add(t)
for {
    if heapQueue.Len() == 0 {
        break
    }
    t = heapQueue.Remove()
    currentNode := t.node
    currentDistance := t.weight
    if currentDistance >
                    distances[convert(currentNode)].weight {
        continue
    }
    for t, w := range graph[currentNode] {
        neighbor := t
        weight := w
        distance := currentDistance + weight
        /*
            Only consider this new path if it's
            better than any path we've already
            found
        */
        if distance <
                distances[convert(neighbor)].weight {
            distances[convert(neighbor)] =
                    tuple{neighbor, distance}
            heapQueue.Add(tuple{neighbor,
                    distance})
        }
```

```go
        }
    }
    return distances
}

func main() {
    graph := make(map[rune]edges)
    graph['A'] = edges{'B': 4, 'H': 1}
    graph['B'] = edges{'A': 4, 'C': 1, 'H': 11}
    graph['C'] = edges{'B': 1, 'I': 2, 'F': 1, 'D': 7}
    graph['D'] = edges{'C': 7, 'F': 8, 'E': 1}
    graph['E'] = edges{'D': 1, 'F': 10}
    graph['F'] = edges{'G': 2, 'C': 1, 'D': 8, 'E': 10}
    graph['G'] = edges{'F': 2, 'I': 1, 'H': 1}
    graph['H'] = edges{'G': 1, 'I': 7, 'B': 11, 'A': 1}
    graph['I'] = edges{'C': 2, 'H': 7, 'G': 1}

    solution := Dijkastra(graph)
    for node, weight := range solution {
        fmt.Printf("%s %d ", string(node + 65), weight)
    }
}

/* Output
A {65 0} B {66 4} C {67 5} D {68 12} E {69 13} F {70 4} G {71 2} H {72 1}
I {73 3}
*/
```

Explanation of Solution

Each node in the graph is represented by a tuple defined as

```go
type tuple struct {
    node rune
    weight int
}

type edges = map[rune]int
```

A **Graph** is defined as:

type Graph map[rune]edges

In a graph, each node, with key of type rune, such as "A", is mapped to another map, edges, from rune to int. This layering of abstractions is needed to represent a structure as complicated as a graph.

A priority queue plays a central role in implementing a solution to the problem. Here, we implement *PriorityQueue* with generic type T by importing package "github.com/jiaxwu/container/heap".

We define a *Less* function that compares the int field of the tuples to order them from smallest to largest.

We initialize the queue using

heapQueue := New[tuple](Less)

Here, we see another example whereby having a generic structure, **PriorityQueue**, makes it easy to create an instance with base-type **tuple**, useful in this application.

We define a slice, **distances**, as containing tuples of node (type rune) and an int that represents the best distance so far. We initialize distances to have very large initial value.

We set **distances[0]** to **tuple{'A', 0}**.

We push this tuple onto the heap queue. This heap queue is set up to ensure that the tuple with the smallest distance is at the head of the line, the first tuple that can be removed.

In a loop that terminates when the queue becomes empty, we remove the head of the queue and assign current distance to its weight field. If this value is greater than the second field of the tuple removed from the queue, we discard it by continuing the loop.

In a second inner loop in which we range over the connections from the current node, we compare the sum of the current distance and the weight of each graph connection to the best distance so far for the given connection. If this best distance so far is less than the value in the distances slice, we push the tuple onto the queue and update the distances slice.

In main, we define the graph shown earlier. The output displays the shortest distances from source node "A" to each of the other nodes.

In the next section, we present another classic graph problem and its solution – minimum spanning tree.

16.5 Minimum Spanning Tree

A **minimum spanning tree** for a weighted undirected connected graph is a collection of edges that touch all nodes of the graph without any cycles and with minimum weight. The weight of the spanning tree is defined as the sum of the weights of the edges that comprise the tree.

Although several people have laid claim to creating an algorithm for creating a minimum spanning tree from a weighted graph, the two most famous algorithms for doing this are by Prim and Kruskal.

We shall present the Kruskal algorithm and its implementation. This algorithm first appeared in *Proceedings of the American Mathematical Society*, pp. 48–50, in 1956 and was written by Joseph Kruskal.

The approach that is taken is to incrementally build the tree one edge at a time by choosing the cheapest edge among those still available. This strategy is a classic greedy strategy in which a local optimization leads to a global optimum solution.

Kruskal Algorithm

1. Sort the edges in ascending order of their weights.

2. Select the edge having minimum weight and add it to the spanning tree providing that a cycle does not occur.

3. Repeat steps 1 and 2 until all nodes have been covered.

To see how the algorithm works, we shall walk through an example. Consider the tree shown in Figure 16-3.

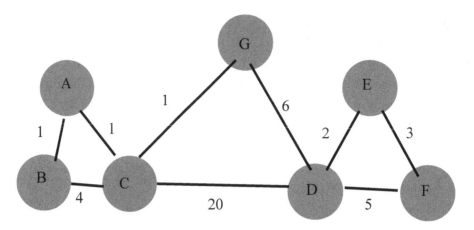

Figure 16-3. *Graph for Kruskal algorithm*

The first edge that we insert into our spanning tree has the smallest weight. There are three such edges, "AB", "AC", and "CG", each with weight equal to 1. We insert all three into the spanning tree since no cycles are produced. Next, we insert "DE" and "EF", with weights 2 and 3, since these produce no cycles.

The next smallest edge is "BC", with a weight of 4. If we were to add this edge to the spanning tree, we would get a cycle among the nodes, "A", "B", and "C". So we reject this edge. Likewise, we reject the next smallest available edge, "DF", since its inclusion would produce a cycle among the nodes, "D", "E", and "F".

The next smallest node, "GD", with weight 6, is added next. The remaining available node, "CD", is rejected since its inclusion would create a cycle among the nodes, "C", "G", and "D".

A minimum spanning tree for this graph is therefore

{1 A B} {1 A C} {1 C G} {2 D E} {3 E F} {6 G D} with a total weight of 14.

In the next section, we implement the Kruskal algorithm. The main driver program uses the graph shown previously.

16.6 Implementation of Kruskal Algorithm

Our implementation of the Kruskal algorithm is relatively short. The details are presented in Listing 16-3 and carefully explained after the listing.

Listing 16-3. Kruskal algorithm

```go
package main

import (
    "fmt"
    "sort"
)

type Edge struct {
    weight int
    node1 Node
    node2 Node
}

type Node = string

type EdgeSlice []Edge

// Infrastructure to allow []Edges to be sorted
func (edges EdgeSlice) Len() int {
    return len(edges)
}

func (edges EdgeSlice) Swap(i, j int) {
    edges[i], edges[j] = edges[j], edges[i]
}

func (edges EdgeSlice) Less(i, j int) bool {
    return edges[i].weight < edges[j].weight
}

var connection map[Node]Node

/*
    The initial level of each Node is 0.
    If the node is node2 of an Edge,
    increase its level by 1.
*/
```

```go
var end map[Node]int

func Initialize(node Node) {
    connection[node] = node
    end[node]  = 0
}

func Find(node Node) Node {
    // Stops a cycle
    if connection[node] != node {
        connection[node] = Find(connection[node])
    }
    return connection[node]
}

func Connect(node1, node2 Node) {
    n1 := Find(node1)
    n2 := Find(node2)
    fmt.Printf("\nFind(%s) = %s", node1, n1)
    fmt.Printf("\nFind(%s) = %s", node2, n2)
    if n1 != n2 {
        fmt.Printf("\nend[%s] = %d", n1, end[n1])
        fmt.Printf("\nend[%s] = %d", n2, end[n2])
        if end[n1] > end[n2] {
            connection[n2] = n1
            fmt.Printf("\nconnection[%s] = %s", n2,
                    n1)
        } else {
            connection[n1] = n2
            fmt.Printf("\nconnection[%s] = %s", n1,
                    n2)
            if end[n1] == end[n2] {
                end[n2] += 1
                fmt.Printf("\nend[%s] = 1", n2)
            }
        }
```

```go
        }
}

func Kruskal(nodes []Node, edges EdgeSlice) []Edge {
    for _, node := range nodes {
        Initialize(node)
    }
    spanningTree := []Edge{}
    sort.Sort(edges)
    for _, edge := range edges {
        node1 := edge.node1
        node2 := edge.node2
        n1 := Find(node1)
        n2 := Find(node2)
        fmt.Printf("\nFind(%s) = %s", node1, n1)
        fmt.Printf("\nFind(%s) = %s", node2, n2)
        if n1 != n2 {
            Connect(node1, node2)
            fmt.Printf("\nConnect(%s, %s)", node1,
                    node2)
            spanningTree = append(spanningTree, edge)
        } else {
            fmt.Printf("\nReject edge %s and %s",
                    node1, node2)
        }
    }
    return spanningTree
}

func main() {
    connection = make(map[Node]Node)
    end = make(map[Node]int)
    // Define the graph by its nodes and edges
    nodes := []Node{"A", "B", "C", "D", "E", "F", "G"}
    edges := []Edge{ {1, "A", "B"}, {1, "A", "C"},
                    {4, "B", "C"}, {20, "C", "D"},
```

```
                    {2, "D", "E"}, {3, "E", "F"},
                    {6, "G", "D"}, {1, "C", "G"},
                    {5, "D", "F"} }
    spanningTree := Kruskal(nodes, edges)
    fmt.Println("\n", spanningTree)
}
/* Output
Find(A) = A
Find(B) = B
Find(A) = A
Find(B) = B
end[A] = 0
end[B] = 0
connection[A] = B
end[B] = 1
Connect(A, B)
Find(A) = B
Find(C) = C
Find(A) = B
Find(C) = C
end[B] = 1
end[C] = 0
connection[C] = B
Connect(A, C)
Find(C) = B
Find(G) = G
Find(C) = B
Find(G) = G
end[B] = 1
end[G] = 0
connection[G] = B
Connect(C, G)
Find(D) = D
Find(E) = E
Find(D) = D
```

```
Find(E) = E
end[D] = 0
end[E] = 0
connection[D] = E
end[E] = 1
Connect(D, E)
Find(E) = E
Find(F) = F
Find(E) = E
Find(F) = F
end[E] = 1
end[F] = 0
connection[F] = E
Connect(E, F)
Find(B) = B
Find(C) = B
Reject edge B and C
Find(D) = E
Find(F) = E
Reject edge D and F
Find(G) = B
Find(D) = E
Find(G) = B
Find(D) = E
end[B] = 1
end[E] = 1
connection[B] = E
end[E] = 1
Connect(G, D)
Find(C) = E
Find(D) = E
Reject edge C and D
 [{1 A B} {1 A C} {1 C G} {2 D E} {3 E F} {6 G D}]*/
```

Explanation of Kruskal Implementation

We walk through the example given in the main driver to uncover the details of this implementation.

We initialize the **connection** and **end** maps.

We define the input, the slice of **Node** values, and the slice of **Edge** values in main.

We pass these slices to the **Kruskal** function.

In function **Kruskal**, we initialize all the nodes by setting the connection of the node to itself and the end value to 0.

We define **spanningTree** as an empty slice of edges. We sort the edges by their weight, having created the infrastructure for ensuring that an edge slice can be sorted (functions **Len**, **Swap**, and **Less**).

In a loop over the edges, for each edge, we define **node1** and **node2** as the beginning and ending nodes of the edge.

The code has been instrumented with many **fmt.Printf** outputs that chronicle the algorithm in detail showing, in particular, the rejection of edges that cause cycles.

Let's take a look.

The first edge that needs to be rejected is the link from "B" to "C". Let's zoom in on the details following the connection from "E" to "F". The next link to be inserted is the link from "C" to "B".

Since Find("B") equals Find("C"), this link is not appended to the spanning tree.

The output lines shown in boldface lead to the rejections of the links shown. The link from "D" to "F" is rejected for the same reason.

16.7 Summary

In this chapter, we introduced graph structures and some applications. We showed several examples of how to represent a graph, and we examined some basic algorithms associated with graph traversal.

The next chapter introduces the famed Travelling Salesperson Problem (TSP). All known exact solutions to this problem are computationally intractable. We introduce one such solution and test it on a smaller-sized problem. We also show how to plot a tour associated with a TSP solution.

CHAPTER 17

Travelling Salesperson Problem

The previous chapter introduced the graph data structure. A generic implementation was shown. Several classic graph algorithms were implemented and discussed.

This chapter is the first of several chapters that examine solutions to the classic Travelling Salesperson Problem. An exact solution to this problem is computationally intractable. In this chapter, we present and implement an algorithm for obtaining an exact solution to this problem.

In the next section, we introduce this classic problem.

17.1 Travelling Salesperson Problem and Its History

The Travelling Salesperson Problem (TSP) is a classic problem with a rich history. Given a set of cities and the distance between every pair of cities, the problem is to **find the shortest tour that visits every city exactly once and returns to the starting city**. The problem was first formulated in 1930. It has become one of the most intensively studied problems in optimization.

Some history:

See

https://en.wikipedia.org/wiki/Travelling_salesman_problem#Exact_
algorithms.

1. An exact solution for 15,112 German towns from TSPLIB was found in 2001 using the **cutting-plane method** proposed by **George Dantzig**, **Ray Fulkerson**, and **Selmer M. Johnson** in 1954, based on **linear programming**.

© Richard Wiener 2022
R. Wiener, *Generic Data Structures and Algorithms in Go*, https://doi.org/10.1007/978-1-4842-8191-8_17

2. In May 2004, the Travelling Salesman Problem of visiting all 24,978 towns in Sweden was solved: a tour of length approximately 72,500 kilometers was computed, and it was proven that no shorter tour exists.

3. In March 2005, the Travelling Salesman Problem of visiting all 33,810 points in a circuit board was solved using **Concorde TSP Solver**. The computation took approximately 15.7 CPU-years.

TSP is a member of a group of problems that are **NP-hard** (nondeterministic polynomial time hard). If a polynomial-time-based solution could be found for any problem in this group, it can be proven that a polynomial-time solution for all the problems in this group could be found. To date, no such polynomial-time solutions have been found for any NP-hard problem.

In the next section, we present a brute-force solution to this problem that produces an exact solution.

17.2 An Exact Brute-Force Solution

Cities will be represented by vertices in a graph and numbered 0, 1, 2, ..., n. The distance between cities will be specified as either integer or floating-point numbers and shown as the edges in the graph.

Consider the graph shown in Figure 17-1. This graph represents a four-city problem. The edge values represent the distance between cities.

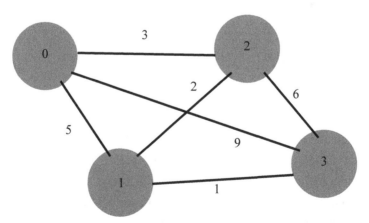

Figure 17-1. *Graph of a four-city TSP*

The brute-force solution requires that we obtain all permutations of tours that start with city 0 and end with city 0. For each tour in the permutation, we compute its cost. We return the tour with the lowest cost.

The tour permutations that we consider are shown in Figure 17-2.

```
[0 1 2 3]
[0 1 3 2]
[0 2 1 3]
[0 2 3 1]
[0 3 2 1]
[0 3 1 2]
```

Figure 17-2. *Tour permutations*

For each tour permutation, we compute the length of the tour. The tour permutation with the smallest length is an optimum solution to the problem. There may be ties.

Finding Permutations

The first task is to compute all the permutations of a slice containing consecutive integers starting at 0.

Listing 17-1 performs this task.

Listing 17-1. Permutations of slice

```go
package main

import (
    "fmt"
)

func Permutations(data []int, operation func([]int)) {
    permute(data, operation, 0)
}

func permute(data []int, operation func([]int), step
            int) {
    if step > len(data) {
        operation(data)
```

```go
        return
    }
    permute(data, operation, step + 1)
    for k := step + 1; k < len(data); k++ {
        data[step], data[k] = data[k], data[step]
        permute(data, operation, step + 1)
        data[step], data[k] = data[k], data[step]
    }
}

func main() {
    data := []int{0, 1, 2, 3}
    Permutations(data, func(a []int) {
        fmt.Println(a)
    })
}
/* Output
[0 1 2 3]
[0 1 3 2]
[0 2 1 3]
[0 2 3 1]
[0 3 2 1]
[0 3 1 2]
[1 0 2 3]
[1 0 3 2]
[1 2 0 3]
[1 2 3 0]
[1 3 2 0]
[1 3 0 2]
[2 1 0 3]
[2 1 3 0]
[2 0 1 3]
[2 0 3 1]
[2 3 0 1]
[2 3 1 0]
[3 1 2 0]
```

```
[3 1 0 2]
[3 2 1 0]
[3 2 0 1]
[3 0 2 1]
[3 0 1 2]
*/
```

We leave it to the reader to walk through the code for a small problem and verify that it produces the desired permutation.

The second parameter in the *Permutations* function is an operation that must be performed on each slice. In the example presented in Listing 17-1, the operation is to output the slice. This is shown in *main* in boldface.

Brute-Force Computation for TSP

Listing 17-2 presents a brute-force computation for the TSP. It uses the permutation logic presented in Listing 17-1. It finds all the permutations of tours that start at 0 and end at 0. For each tour, it computes the cost of the tour and saves the best tour along with its cost.

Listing 17-2. Brute-force solution to TSP

```go
package main

import (
    "fmt"
    "math/rand"
    "time"
)

type Graph [][]int

type TourCost struct {
    cost int
    tour []int
}

var minimumTourCost TourCost
var graph Graph
```

```
func Permutations(data []int, operation func([]int)) {
    permute(data, operation, 0)
}

func permute(data []int, operation func([]int), step
            int) {
    if step > len(data) {
        operation(data)
        return
    }
    permute(data, operation, step+1)
    for k := step + 1; k < len(data); k++ {
        data[step], data[k] = data[k], data[step]
        permute(data, operation, step+1)
        data[step], data[k] = data[k], data[step]
    }
}

func TSP(graph Graph, numCities int) {
    tour := []int{}
    for i := 1; i < numCities; i++ {
        tour = append(tour, i)
    }
    minimumTourCost = TourCost{32767, []int{}}
    Permutations(tour, func(tour []int) {
        // Compute cost of tour
        cost := graph[0][tour[0]]
        for i := 0; i < len(tour)-1; i++ {
            cost += graph[tour[i]][tour[i+1]]
        }
        cost += graph[tour[len(tour)-1]][0]
        if cost < minimumTourCost.cost {
            minimumTourCost.cost = cost
            var tourCopy []int
            tourCopy = append(tourCopy, 0)
            tourCopy = append(tourCopy, tour...)
```

```go
            tourCopy = append(tourCopy, 0)

            minimumTourCost.tour = tourCopy
        }
    })
}

func main() {
    graph = Graph{{0, 5, 3, 9}, {5, 0, 2, 1}, {3, 2, 0, 6},
                {9, 1, 6, 0}}
    TSP(graph, 4)
    fmt.Printf("\nOptimum tour cost: %d  An Optimum
            Tour %v", minimumTourCost.cost,
            minimumTourCost.tour)

    numCities := 14
    graph2 := make([][]int, numCities)
    for i := 0; i < numCities; i++ {
        graph2[i] = make([]int, numCities)
    }
    for row := 0; row < numCities; row++ {
        for col := 0; col < numCities; col++ {
            graph2[row][col] = rand.Intn(9) + 2
        }
    }

    // Create a short path for test purposes
    for i := 0; i < numCities-1; i++ {
        graph2[i][i+1] = 1
    }
    graph2[numCities-1][0] = 1

    start := time.Now()
    TSP(graph2, numCities)
    elapsed := time.Since(start)
    fmt.Printf("\nOptimum tour cost: %d  An Optimum
            Tour %v", minimumTourCost.cost,
              minimumTourCost.tour)
```

```
      fmt.Println("\nComputation time: ", elapsed)
}
/* Output
Optimum tour cost: 15  An Optimum Tour [0 1 3 2 0]
Optimum tour cost: 14  An Optimum Tour [0 1 2 3 4 5 6 7 8 9 10 11 12 13 0]
Computation time:   2m15.918717943s
*/
```

Discussion of Code

We focus on the invocation of **_Permutations_** inside of function **_TSP_**. In particular, we look at the function defined as the second parameter, shown in boldface.

For each tour in the permutation, we compute the cost of the tour.

The first cost computed is the cost of going from city 0 to the first city in the tour permutation. Following that, in a loop, we compute and add the costs for the sequence of cities in the tour permutation. The final cost computed is the cost from the last city in the tour permutation back to city 0.

We compare the cost of the tour permutation with the lowest cost thus far. This is held as a global variable of type **_TourCost_**.

```
type TourCost struct {
    cost int
    tour []int
}
```

A programming subtlety requires that we make a copy of the tour that we save in the global **_minimumTourCost_**. This is needed because assigning one slice to another makes a shallow copy. We are interested here in copying the information, not the address of the slice.

We accomplish this with the append function as follows:

```
var tourCopy []int
tourCopy = append(tourCopy, 0)
tourCopy = append(tourCopy, tour...)
tourCopy = append(tourCopy, 0)
```

This block of code also adds the tour link from the starting city 0 and the tour link of getting back to city 0.

The computational cost of this brute-force solution is **(n – 1)**! That is why the brute-force method is intractable.

To illustrate this, we solve a 14-city problem with random integer distances between cities. We embed a low-cost path from city 0 to 1, 1 to 2, …, 13 to 0, each a distance 1 apart, for a total cost of 14. This does not affect computation time but gives us a test of correctness of the TSP algorithm.

As you can see from the output, we pass this test. The computation time for a 14-city problem is over two minutes.

If we were to increase the size of the problem by one city, the computation time would increase by a factor of 14. The computational complexity, O(n!), clearly makes this brute-force algorithm intractable.

Other Solutions

There are many algorithms, all intractable, that produce exact solutions to TSP. These algorithms employ dynamic programming, branch and bound, linear programming, and other techniques. They work well for small-sized problems but are impractical when the number of cities exceeds several dozen.

Before we examine heuristic algorithms for solving TSP that produce solutions close to the exact solution in reasonable time and storage space for large-sized problems, the next section presents code for displaying a TSP tour.

17.3 Displaying a TSP Tour

Listing 17-3 uses a third-party package to graphically display a tour given a slice of points that define the cities in the tour.

Listing 17-3. Displaying a TSP tour

```
package main

import (
    "image/color"
    "gonum.org/v1/plot"
```

```go
    "gonum.org/v1/plot/plotter"
    "gonum.org/v1/plot/vg"
    "gonum.org/v1/plot/vg/draw"
)

type Point struct {
    X float64
    Y float64
}

func definePoints(cities []Point, tour []int)
                  plotter.XYs {
    pts := make(plotter.XYs, len(cities) + 1)
    pts[0].X = cities[0].X
    pts[0].Y = cities[0].Y
    for i := 1; i < len(cities); i++ {
        pts[i].X = cities[tour[i]].X
        pts[i].Y = cities[tour[i]].Y
    }
    pts[len(cities)].X = cities[0].X
    pts[len(cities)].Y = cities[0].Y
    return pts
}

func DrawTour(cities []Point, tour []int) {
    data := definePoints(cities, tour) // plotter.XYs
    p := plot.New()
    p.Title.Text = "TSP Tour"
    lines, points, err := plotter.NewLinePoints(data)
    if err != nil {
        panic(err)
    }
    lines.Color = color.RGBA{R: 255, A: 255}
    points.Shape = draw.PyramidGlyph{}
    points.Color = color.RGBA{B: 255, A: 255}
    p.Add(lines, points)
    // Save the plot to a PNG file.
```

```go
    if err := p.Save(4*vg.Inch, 4*vg.Inch,
            "tour.png"); err != nil {
        panic(err)
    }
}

func main() {
    numCities := 4
    cities := make([]Point, numCities)
    cities[0] = Point{0.0, 0.0}
    cities[1] = Point{3.0, 0.0}
    cities[2] = Point{3.0, 4.0}
    cities[3] = Point{1.0, 11.0}
    tour := []int{0, 3, 1, 2}
    DrawTour(cities, tour)
}
```

Discussion of Code

The helper function, ***definePoints***, returns *a plotter.XYs*. It uses the input slice, ***cities***, to obtain the **X** and **Y** coordinates of each city that are assigned to ***pts***. It assigns the sequence of points based on the input slice, ***tour***.

The ***DrawTour*** function invokes ***definePoints*** and assigns the result to ***data***. The remaining code follows the protocol in the plot package that is imported. A new plot, ***p***, is defined. The ***lines*** and ***points*** variables are obtained from ***plotter.NewLinePoints***.

After adding these to the plot, ***p***, a **png** file is saved, which contains the points and lines that graphically display the tour.

The output of this program is shown in Figure 17-3.

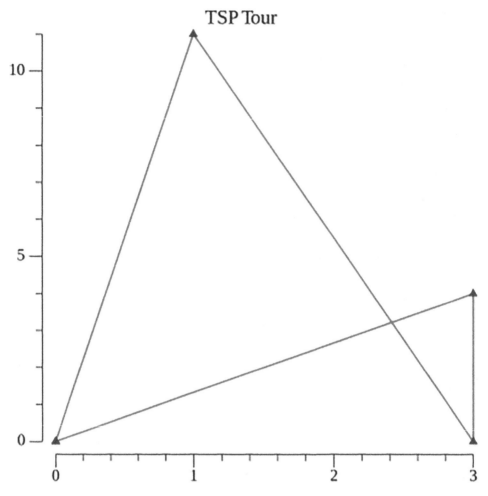

Figure 17-3. *Output from Listing 17-3*

17.4 Summary

The famed Travelling Salesperson Problem is introduced. A brute-force solution is presented. This solution, like all known exact solutions, is computationally intractable with a big O of **O(n!)**.

Code for displaying a tour with specified coordinate locations is presented and illustrated with a simple example.

In the next chapter, we present another algorithm for obtaining an exact solution solving TSP. This algorithm uses a powerful technique called branch and bound. Like all known exact solutions to TSP, this branch-and-bound algorithm is computationally intractable.

CHAPTER 18

Branch-and-Bound Solution to TSP

The previous chapter introduced the famed Travelling Salesperson Problem (TSP). A brute-force solution was presented. Like all exact solutions to this problem, it is computationally intractable. A third-party package was presented along with code for graphically displaying a TSP tour.

This chapter presents another exact solution to TSP using a powerful technique, branch and bound. It too is computationally intractable.

In the next section, we introduce the technique of branch and bound and show how it can be applied to TSP.

18.1 Branch and Bound for TSP

Much of this chapter is based on a paper written by this author and published in the *Journal of Object Technology* in 2003. The paper is "Branch and Bound Implementation for the Traveling Salesperson Problem" (Richard Wiener, Journal of Object Technology, Vol 2. No. 3, May–June 2003).

We are given a graph that contains distances connecting all cities (each city is connected to every other city with an edge representing the distance between the two cities). The nodes of the graph are the cities, numbered 0, ..., n - 1. A tour is a sequence of cities that starts at city 0, visits each of the other cities exactly once, and returns to the starting city 0.

In any tour, the value of an edge when leaving a city must be equal or greater than the value of the shortest edge leaving the city.

This forms the basis for a branch-and-bound solution to TSP.

© Richard Wiener 2022
R. Wiener, *Generic Data Structures and Algorithms in Go*, https://doi.org/10.1007/978-1-4842-8191-8_18

An Example

Consider a TSP with the following distance matrix (cost matrix):

```
 0 14  4 11 20
14  0  7  8  7
 4  7  0  7 16
11  8  7  0  2
20  7 16  2  0
```

We build a solution tree as follows:

At level 0, the root node represents a partial tour [0].

At level 1, nodes representing the partial tours [0, 1], [0, 2], [0, 3], ..., [0, n - 1] are generated.

At level 2, nodes representing the partial tours [0, 1, 2], [0, 1, 3], [0, 1, 4], ... are generated.

This pattern continues until at the lowest level, we have every permutation for all tours.

Computation of Lower Bound

For the root node of the preceding graph, with partial tour [0], the lower bound on the costs of leaving the five vertices is

City 0: minimum (14, 4, 11, 20) = **4**.

City 1: minimum (14, 7, 8, 7) = **7**.

City 2: minimum (4, 7, 7, 16) = **4**.

City 3: minimum (11, 8, 7, 2) = **2**.

City 4: minimum (20, 7, 16, 2) = **2**.

Therefore, the lower bound for the TSP solution based on the partial tour [0] is the sum of these values, which is **19**.

Let us compute the lower bound for a partial tour [0, 2, 3].

City 0: **4** (since the tour contains 0 -> 2)

City 1: **7** (cannot touch node already on tour)

City 2: **7** (since the tour contains 2 -> 3)

City 3: **11** (since the tour contains 3 -> 0)

City 4: **7** (cannot touch node already on tour)

Therefore, the lower bound for the partial tour [0, 2, 3] is the sum, which equals **36**.

The computational cost of computing the lower bound for a partial tour is low.

At any level in the tree containing partial tours, the nodes can be ranked by their computed lower bounds. We can use a priority queue to hold the tree structure.

Branch-and-Bound Algorithm

- Set an initial value for the best tour cost.

- Initialize a priority queue (PQ).

- Generate the first node with partial tour [0] and compute its lower bound.

- Insert this node into PQ.

- While the PQ is not empty, remove the first node from the PQ and assign it to parent.

- If its lower bound < best tour cost, set its level to the level of parent node + 1.

- If this level is N – 1, where N is the number of cities, add starting city 0 to the end of the partial tour and compute the cost of the full tour.

- If this cost of full tour < best tour cost, update the best tour cost and save the best tour.

- If the level of the parent node + 1 is not equal to N – 1,

- For all i such that 1 <= i < N and i is not in the partial tour of the parent,

- Copy the partial tour from parent to new node and add i to the end of this partial tour.

- Compute the lower bound for this new node.

- If this lower bound is less than the best tour cost, insert this new node into the priority queue; otherwise, prune this node.

The Priority Queue

Before a node is inserted into the PQ, it is screened to determine whether its lower bound is less than the currently known best tour. This helps to keep the number of nodes in the PQ to a manageable level.

What priority rules must the queue enforce?

Nodes at a deeper level (higher level number) have priority over nodes at a shallower level in the PQ. This assures that the tree grows downward and that leaf nodes representing complete tours are generated as quickly as possible. In comparing two nodes at the same level, priority is given to the node with the smallest lower bound. In the event of a tie (two nodes with equal lower bound), the sum of the cities in the partial tour is computed. The node with the smaller sum is given a higher priority than the node with the larger sum (a tie cannot occur). The rules just stated disallow two distinct nodes from having an equal priority.

We walk through a portion of the five-city example presented earlier to see how the priority queue is built.

A Walk-Through Part of the Five-City Example Presented Earlier

The initial cost is computed by finding the cost of the tour, [0, 1, ..., n – 1, 0], which is **50**. Since the lower bound (LB) of the root node was shown earlier to be 19, we push the partial tour [0] onto the PQ. We remove tour[0] from the PQ and generate nodes at level 1. All but one of these nodes have LB < 50, so we push them onto the PQ. The top of the PQ is [0, 2] with LB = 19. Next comes [0, 1] with LB = 39, and third comes [0, 3] with LB= 43. We continue to generate nodes as specified in the algorithm and as shown in Figure 18-1.

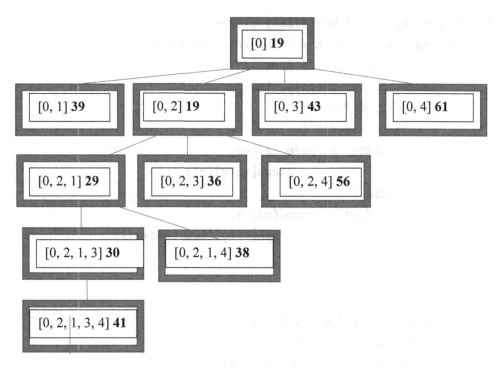

Figure 18-1. *Part of solution tree*

The full tour [0, 2, 1, 3, 4] is the first to be generated. Since its cost is less than the best so far of 50, we assign the best tour to be [0, 2, 1, 3, 4] and its cost 41.

The front of the PQ contains the node [0, 2, 1, 4] since it is at the deepest level. The algorithm backtracks to that node and then generates another full tour [0, 2, 1, 4, 3] also with cost 41. This allows some of the nodes to be pruned when they are taken off the PQ since their lower bounds are greater than **41**.

You may wish to continue this process for this example.

In the next section, we look at the implementation of this algorithm.

18.2 Branch-and-Bound Implementation

One of the most important functions needed to support this branch-and-bound algorithm is the computation of a lower bound for a given tour. This function is as follows:

```
func LowerBound(tour []int) float64 {
    edges := make([]float64, 0)
    sum := 0.0
    n := len(tour)
```

```
    for city := 0; city < NUMCITIES; city++ {
        for index := 0; index < NUMCITIES; index++ {
            // index is part of tour
            found, pos := In(city, tour)
            if n > 1 && found {
                if pos == n-1 {
                    edges = append(edges,
                                graph[city][0])
                } else {
                    edges = append(edges,
                                graph[city][tour[pos+1]])
                }
                break
            }
            found, _ = In(index, tour)
            if n == 1 || !found {
                // Don't allow an index already in
                // tour
                edges = append(edges,
                            graph[city][index])
            }
        }
        sum += Minimum(edges)
        edges = make([]float64, 0)
    }
    return sum
}
```

The function works exactly as specified in the outline of the algorithm presented in Section 18.1.

A key support function is *In* given as follows:

```
func In(value int, values []int) (bool, int) {
    // Returns true if value in values
    // Returns index of location or -1 if not found
    for index := 0; index < len(values); index++ {
```

```
        if values[index] == value {
            return true, index
        }
    }
    return false, -1
}
```

Implementation of Priority Queue

The priority queue plays a central role in this algorithm. It stores nodes, each containing a tour, a lower bound, and a level with priorities defined as specified in Section 18.1 and repeated here for your convenience:

- Nodes at a deeper level (higher level number) have priority over nodes at a shallower level in the PQ.

- Priority is given to the node with the smallest lower bound if the nodes are at the same level.

- If two nodes at the same level have the same lower bound, we add up the cities in the node's tour, and the node with the higher sum has a higher priority.

The code that supports the implementation of the priority queue is given as follows:

```
type Node struct {
    tour       []int
    lowerBound float64
    level      int
}

type Nodes []Node

// Allow nodes to be sorted
func (nodes Nodes) Len() int {
    return len(nodes)
}
```

```go
func (nodes Nodes) Swap(i, j int) {
    nodes[i], nodes[j] = nodes[j], nodes[i]
}

func (nodes Nodes) Less(i, j int) bool {
    if nodes[i].level > nodes[j].level {
        return true
    }
    if nodes[i].level == nodes[j].level &&
        nodes[i].lowerBound == nodes[j].lowerBound {
        // Return the smaller sum of cities
        tour1 := nodes[i].tour
        sum1 := 0;
        for i := 0; i < len(tour1); i++ {
            sum1 += tour1[i]
        }
        tour2 := nodes[j].tour
        sum2 := 0;
        for i := 0; i < len(tour2); i++ {
            sum2 += tour2[i]
        }
        return sum1 < sum2
    }
    if nodes[i].level == nodes[j].level &&
        nodes[i].lowerBound != nodes[j].lowerBound {
        return nodes[i].lowerBound <
                    nodes[j].lowerBound
    }
    return false
}

type PriorityQueue struct {
    items Nodes
}
func NewPriorityQueue() PriorityQueue {
    return PriorityQueue{}
}
```

```go
func (pq *PriorityQueue) Insert(node Node) {
    tourToInsert := DeepCopy(node.tour)
    nodeToInsert := Node{tourToInsert, node.lowerBound,
                            node.level}
    pq.items = append(pq.items, nodeToInsert)
    sort.Sort(pq.items)
}
func (pq *PriorityQueue) Remove() Node {
    result := pq.items[0]
    pq.items = pq.items[1:]
    return result
}
```

The priority queue (PQ) holds entities of type Node. Each time we insert a new node into the queue, we sort the queue to ensure that the node with the highest priority is at the front of the queue.

To enable the sorting of nodes in the PQ, we need to implement the interface to the **Sort** method from package **sort**.

This is accomplished with the functions *Len*, *Swap*, and *Less* as given earlier. Each of the three priority rules is implemented in the function *Less*.

The *Insert* function sorts the items (slice of *Node*) after appending the node being inserted.

Generating Branch-and-Bound Solution

The function that generates nodes according to the outline from Section 18.1, inserts them into the priority queue, finds best new tours, and backtracks while pruning nodes whose lower bound exceeds the known best tour to date is given as follows in function TSP:

```go
func TSP() {
    var elapsed time.Duration
    start := time.Now()
    bestTour := []int{}
    for i := 0; i < NUMCITIES; i++ {
        bestTour = append(bestTour, i)
    }
```

```go
pq := NewPriorityQueue()
bestCost := LowerBound(bestTour)
tour := []int{0}
lowerBound := LowerBound(tour)
node := Node{tour, lowerBound, 0}
nodesGenerated += 1
pq.Insert(node)
for {
    if len(pq.items) == 0 {
        break
    }
    top := pq.Remove()
    topLevel := top.level
    topTour := top.tour

    // Generate nodes at topLevel + 1
    for i := 0; i < NUMCITIES; i++ {
        tour := DeepCopy(topTour)
        found, _ := In(i, topTour)
        if !found {
            tour = append(tour, i)
            nodesGenerated += 1
            if nodesGenerated %
                    10_000_000 == 0 {
                fmt.Println("\nNodes generated (in
                millions): ", nodesGenerated /
                                1_000_000)
                fmt.Printf("\n\nOptimum tour cost:
                %0.2f  \nBest tour: %v", bestCost,
                bestTour)
                elapsed = time.Since(start)
                seconds := elapsed / 1_000_000_000
                rate := float64(nodesGenerated) /
                        float64(seconds)
                fmt.Printf("\nNodes generated per
                second: %0.0f  Length of PQ: %d
```

```
                Time elapsed: %v", rate,
                len(pq.items), elapsed)
        }
        if len(tour) == NUMCITIES {
            // A complete tour is obtained
            tourCost := LowerBound(tour)
            if tourCost < bestCost {
                bestTour = tour
                bestCost = tourCost
                fmt.Println("\n\nBest cost of
                tour so far: ", bestCost)
            }
        } else {
            tourCost := LowerBound(tour)
            if tourCost < bestCost {
                node := Node{tour, tourCost,
                            topLevel + 1}
                pq.Insert(node)
            }
        }
    }
}
fmt.Printf("\n\nOptimum tour cost: %0.2f  \nBest
tour: %v  \nNodes generated: %d", bestCost,
bestTour, nodesGenerated)
}
```

Listing 18-1 puts all the pieces together and shows all the support functions and a main driver that attempts to solve a 33-city problem.

Listing 18-1. Branch-and-bound solution to TSP

```go
package main

import (
    "fmt"
    "sort"
    "time"
)

const (
    NUMCITIES = 33
)

type Node struct {
    tour        []int
    lowerBound  float64
    level       int
}
type Graph [][]float64

var graph Graph
var nodesGenerated int64

type Nodes []Node

// Allow nodes to be sorted
func (nodes Nodes) Len() int {
    return len(nodes)
}

func (nodes Nodes) Swap(i, j int) {
    nodes[i], nodes[j] = nodes[j], nodes[i]
}

func (nodes Nodes) Less(i, j int) bool {
    // Snip
}
```

```go
type PriorityQueue struct {
    items Nodes
}
func NewPriorityQueue() PriorityQueue {
    return PriorityQueue{}
}

func (pq *PriorityQueue) Insert(node Node) {
    // Snip
}

func (pq *PriorityQueue) Remove() Node {
    // Snip
}

func DeepCopy(tour []int) []int {
    result := []int{}
    for i := range tour {
        result = append(result, tour[i])
    }
    return result
}

func Minimum(values []float64) float64 {
    // This function excludes value 0
    min := 100000000.0
    for i := 0; i < len(values); i++ {
        if values[i] != 0 && values[i] < min {
            min = values[i]
        }
    }
    if min == 100000000.0 {
        return 0.0
    }
    return min
}
```

```
func In(value int, values []int) (bool, int) {
    // Snip
}

func LowerBound(tour []int) float64 {
    // Snip
}

func TSP() {
    // Snip
}

func main() {
    graph = // Download from publisher's website
    TSP()
}
```

Data for main

The graph is constructed from data taken from a Rand McNally Atlas of 33 US cities with distances between them specified.

The details in ***main*** are omitted here because of limited space. Please download the entire listing from the publisher's website.

The known solution to this problem is an optimum tour of **10,861** miles.

Results

After running the branch-and-bound program for **18 hours** and 42 minutes and generating **4.6 billion nodes** at about **70,000 nodes per second**, the best tour generated so far is

[0 13 1 2 3 5 4 6 7 8 9 10 11 17 18 19 27 29 30 28 31 32 22 21 23 24 25 26 20 14 15 16 12]

This tour is 11,553 **miles**, which is an error of about 6 percent. We must remind ourselves that the total number of nodes in the full tree representing this problem is 26 3,130,836,933,693,530,167,218,012,160,000,000 nodes.

At an average rate of 71,874 nodes per second, it would take about 1.39×10^{30} seconds or about 4.41×10^{22} years to generate all the leaf nodes in this tree.

18.3 Summary

We presented a branch-and-bound algorithm that provides an exact solution to the TSP. The problem is that this is a computationally intractable problem. If allowed to run long enough, it will find an exact solution to the problem.

A priority queue is used to hold a sequence of nodes, where each node contains a tour (may be a partial tour), a lower bound, and a level. The nodes are sorted in the queue by level, lower bound, and, in the case of a tie, sum of city values in the tour.

A 33-city problem is attempted. After two hours of computation, the best tour obtained is about 6 percent higher than the known optimum tour. This best tour so far remains the same after 16 hours.

This sets the stage for the next two chapters in which we present heuristic solutions to the TSP. These heuristic solutions execute in a matter of seconds and produce solutions that are either exact or have small error.

The next chapter presents a heuristic algorithm, simulated annealing.

CHAPTER 19

Simulated Annealing Heuristic Solution to TSP

The previous chapter presented a branch-and-bound algorithm for producing an exact solution to TSP. Like all known exact solutions, it is computationally intractable.

This chapter presents the powerful simulated annealing heuristic solution to the TSP.

In the next section, we introduce combinatorial optimization problems and set the stage for our presentation of heuristic algorithms for solving TSP.

19.1 Combinatorial Optimization

Combinatorial optimization problems are computationally hard. As the size of these problems increases, the computation time for an exact solution becomes infeasible – years or centuries of computation time and memory requirements that are not realizable.

One such famous and interesting problem is the ***Travelling Salesperson Problem*** (TSP), introduced in the previous chapter. The problem is easy to state and understand. Given n cities, with specified distances between the cities, find the shortest tour that starts at a given city, say, city 0; visits the remaining n – 1 cities; and returns to the starting city. In other words, find the sequence of cities visited so that the total distance travelled is minimum.

Often, we are given the location of the n cities from which we can compute distances between them. We shall assume bidirectional links between cities in a fully connected graph (every city has a link to every other city).

As we saw in Chapter 17, a brute-force solution that enumerates all possible combinations of tours and chooses the tour of lowest distance travelled is of complexity (n – 1)! For example, for four cities labelled 1,2 3, and 4, the possible tours starting and returning to city 1 are

© Richard Wiener 2022
R. Wiener, *Generic Data Structures and Algorithms in Go*, https://doi.org/10.1007/978-1-4842-8191-8_19

```
1 -> 2 -> 3 -> 4
1 -> 2 -> 4 -> 3
1 -> 3 -> 2 -> 4
1 -> 3 -> 4 -> 2
1 -> 4 -> 2 -> 3
1 -> 4 -> 3 -> 2
```

As predicted, there are (4 – 1)! or 3! = 6 possible tours.

For a 29-city problem, the number of possible tours is 28! = 3.0488834e+29. If we could evaluate the tours at a rate of 10 million tours per second, it would take about 317,000,000,000,000 years to complete this computation.

So getting an exact solution to TSP is computationally intractable for problems even of modest size such as a 29-city problem.

Heuristic Solutions

Since there are many important "real-world" applications of TSP (e.g., printed circuit boards, transportation), researchers have devised many heuristic solutions to TSP. A heuristic solution is one that is not guaranteed to be optimum but is computationally tractable (polynomial complexity) and hopefully produces a solution near optimal. Often. a heuristic algorithm will produce the optimum solution to the problem.

Two major frameworks for such approximate solutions to the TSP are

1.   Simulated annealing

2.   Genetic programming

What makes the two major frameworks given previously so interesting is that they are each taken from processes and systems unrelated to TSP specifically. They can be deployed on a wide range of combinatorial optimizations problems. There are other such frameworks that are used to obtain heuristic solutions to TSP.

The first of these heuristic frameworks utilizes models from thermodynamics and the second from genetics and survival of the fittest.

In the next section, we examine the first of these heuristic frameworks, simulated annealing.

19.2 Simulated Annealing

The seminal work that has led to this framework for solving combinatorial optimization problems was published in 1953 by Nicholas Constantine Metropolis. Later, he and a colleague, W.K. Hastings, published the Metropolis-Hastings algorithm, which forms the basis of simulated annealing.

Simulated annealing is a Monte Carlo algorithm. Such algorithms rely on repeated statistical sampling of a system. Many random configurations of the system are generated while computing properties of interest and refining the sampling based on experimental results.

Simulated annealing relies on mimicking the thermodynamic properties of the molecular lattice structure of metal as it is heated and slowly cooled to produce a rigid and strong lattice structure. This process when done by metallurgists is called "annealing." It is done to minimize microscopic deformities in steel load-bearing beams by creating a lattice structure of minimum internal energy. The average energy in such beams during the annealing process (slow cooling of the beam) is given by the Boltzmann factor, $e^{-E/kT}$, where E is the average energy of the beam, T is the temperature, and k is the Boltzmann constant.

A critically important part of the physical annealing process of metallic beams is to lower the temperature slowly. This increases the probability that the internal lattice structure of the beam has minimum energy and is strongest.

Simulated Annealing Steps

An outline of the simulated-annealing algorithm is the following:

1. Choose an initial tour and find its cost.

2. Choose a high initial temperature T for an artificial temperature variable.

3. Modify the tour by making a change to the existing tour (e.g., modifying the order of two cities in the tour or other modifications to be seen shortly).

4. If the new tour cost is smaller than the old tour cost, accept this new tour (downhill move).

5. If the new tour cost is higher than the old tour cost, accept this uphill move with probability given by the Boltzmann factor, $e^{-E/kT}$.

6. At high temperature, the probability of accepting such an "uphill move" is close to 1. This allows the simulation to explore many regions in the solution space and not be driven into a local valley in the tour-cost vs. temperature space.

7. Lower the temperature based on a cooling curve. This cooling curve can be obtained empirically by observing the rate of decline in tour cost as a function of temperature and slowing down the reduction in temperature when this rate of decline is high.

8. Repeat steps 3 to 7.

9. As the temperature gets lower, the probability of accepting uphill moves decreases. This allows a descent hopefully to a lowest energy state (lowest tour cost) close to the global minimum.

What is so special about this algorithm is that it evolves statistically to better and better tours as the temperature variable is slowly lowered. It is a guided random walk of the solution space based on the physics of metallurgy annealing.

It is relatively easy to implement and, as we will demonstrate, produces high-quality solutions, either optimum or close to optimum.

Problem of Convergence to Local Minimum Rather Than Global Minimum

An ever-present challenge in solving combinatorial optimization problems is having a solution converge to a local minimum rather than the desired global minimum. For this reason, it is desirable to allow the solution space to be explored and not to be in a great rush to evolve to a solution.

In the simulated annealing algorithm, we achieve this by allowing "uphill moves." These are moves in the solution space that produce tours that are greater than the best-known tour to date. The goal is to be able to climb out of local valleys in the solution space while finding deeper valleys that hopefully contain the global minimum.

In the next section, we present an implementation of simulated annealing that follows the steps just presented.

19.3 Implementation of Simulated Annealing

We create a type **Status**, which encapsulates the relevant state information about the system.

```
type Status struct {
    tour            []int
    bestTour        []int
    bestCostToDate  float64
    previousCost    float64
    temperature     float64
    downhillMoves   int
    uphillMoves     int
    rejectedMoves   int
    inverseOps      int
    swapOps         int
    insertOps       int
}

var status Status
```

There are three separate operations that we use to perturb a tour in the solution space:

- **Inverse Operation** – We reverse the tour within two index values chosen randomly.

- **Swap** – We swap two cities, chosen randomly, in a given tour.

- **Insert** – Move city in random position second to random position first.

The details of this simulated annealing implementation along with extensive comments and program output are presented in Listing 19-1.

Listing 19-1. Simulated annealing solution to TSP

```go
package main

import (
    "fmt"
    "math"
    "math/rand"
    "time"
)

const (
    NUMCITIES = 29
)

type Point struct {
    x float64
    y float64
}

func init() {
    rand.Seed(time.Now().UnixNano())
}

func (pt Point) distance(other Point) float64 {
    dx := pt.x - other.x
    dy := pt.y - other.y
    return math.Sqrt(dx*dx + dy*dy)
}

func createGraph(numCities int, cities []Point, graph
                    [][]float64) {
    for row := 0; row < numCities; row++ {
        for col := 0; col < numCities; col++ {
            if row == col {
                graph[row][col] = 0.0
            } else {
                graph[row][col] =
                    cities[row].distance(cities[col])
```

```go
            }
        }
    }
}

func cost(graph [][]float64, tour []int) float64 {
    result := 0.0
    for index := 0; index < len(tour) - 2; index++ {
        result += graph[tour[index]][tour[index+1]]
    }
    result += graph[tour[NUMCITIES - 1]][tour[0]]
    return result
}

func randomFrom(min int, max int) int {
    return rand.Intn(max - min) + min
}

func inverseOperation(tour []int) []int {
    /*
        Choose city i randomly from 1 to count - 1.
        Choose city j randomly from 1 to count - 1
        let first be the minimum of index i and j.
        let second be the larger of index i and j.
        reverse the order of cities in the tour from
        index first to index second
        Consider tour = [0, 3, 2, 1, 5, 4] and first = 1
        and second = 4
        The segment 3, 2, 1, 5 is replaced by 5, 1, 2, 3
        and the new tour is
        [0, 5, 1, 2, 3, 4].
    */
    // Choose first and second
    firstIndex := randomFrom(1, len(tour) - 1)
    secondIndex := randomFrom(1, len(tour) - 1)
    for firstIndex == secondIndex {
        firstIndex = randomFrom(1, len(tour) - 1)
```

```go
        secondIndex = randomFrom(1, len(tour) - 1)
    }
    if firstIndex > secondIndex {
        firstIndex, secondIndex = secondIndex,
            firstIndex
    }
    result := deepcopy(tour[:firstIndex])
    for index := 0; index <
        (secondIndex - firstIndex + 1); index += 1 {
        result = append(result, tour[secondIndex -
                                        index])
    }
    for index := secondIndex + 1; index < len(tour);
            index += 1 {
        result = append(result, tour[index])
    }
    return result
}

func swap(tour []int) []int {
    /*
        Swap the city in position first with city in
        position second
        Consider tour [0, 3, 2, 1, 5, 4] and first = 1
        and second = 4
        The new tour would be [0, 5, 2, 1, 3, 4]
    */

    // Choose first and second
    firstIndex := randomFrom(1, len(tour) -  1)
    secondIndex := randomFrom(1, len(tour) - 1)
    for firstIndex == secondIndex {
        firstIndex = randomFrom(1, len(tour) - 1)
        secondIndex = randomFrom(1, len(tour) - 1)
    }
```

```go
    if firstIndex > secondIndex {
        firstIndex, secondIndex = secondIndex,
            firstIndex
    }
    result := deepcopy(tour)
    result[firstIndex], result[secondIndex] =
            result[secondIndex], result[firstIndex]
    return result
}

func insert(tour []int) []int {
    /*
        It means to move the city in position second to
        position first.
        Consider tour [0, 3, 2, 1, 5, 4] and first = 1
        and second = 4
        The new tour would be [0, 5, 3, 2, 1, 4]
    */

    // Choose first and second
    // Choose first and second
    firstIndex := randomFrom(1, len(tour) - 1)
    secondIndex := randomFrom(1, len(tour) - 1)
    for firstIndex == secondIndex {
        firstIndex = randomFrom(1, len(tour) - 1)
        secondIndex = randomFrom(1, len(tour) - 1)
    }
    if firstIndex > secondIndex {
        firstIndex, secondIndex = secondIndex,
            firstIndex
    }
    result := []int{}
    for index := 0; index < len(tour) + 1; index += 1 {
        if index < firstIndex {
            result = append(result, tour[index])
        } else if index == firstIndex {
```

```go
            result = append(result, tour[secondIndex])
        } else if index > firstIndex && index !=
                        secondIndex + 1 {
            result = append(result, tour[index-1])
        }
    }
    return result
}

type Status struct {
    tour          []int
    bestTour      []int
    bestCostToDate float64
    previousCost   float64
    temperature    float64
    downhillMoves  int
    uphillMoves    int
    rejectedMoves  int
    inverseOps     int
    swapOps        int
    insertOps      int
}

var status Status

func deepcopy(tour []int) []int {
    result := []int{}
    for i := range tour {
        result = append(result, tour[i])
    }
    return result
}

func simulatedAnnealing(graph [][]float64) {
    for i := 0; i < NUMCITIES; i++ {
        status.tour = append(status.tour, i)
    }
```

```go
status.tour = append(status.tour, 0)
fmt.Printf("\n\nCost of initial tour %v is %f\n\n",
        status.tour, cost(graph, status.tour))
status.bestTour = deepcopy(status.tour)
status.bestCostToDate = cost(graph,
                             status.bestTour)
status.previousCost = status.bestCostToDate
numberIterationsAtTemperature := 5000
lowestTemperature := 5.0
for status.temperature >= lowestTemperature {
    for iteration := 0; iteration <
            numberIterationsAtTemperature; iteration
                += 1 {
        tour1 := inverseOperation(status.tour)
        cost1 := cost(graph, tour1)
        tour2 := swap(status.tour)
        cost2 := cost(graph, tour2)
        tour3 := insert(status.tour)
        cost3 := cost(graph, tour3)
        newCost1 := math.Min(cost1, cost2)
        newCost := math.Min(newCost1, cost3)

        if newCost == cost1 {
            status.inverseOps += 1
            // Determine whether to accept this
            // tour1
            if newCost < status.previousCost {
                status.downhillMoves += 1
                status.previousCost = newCost
                status.tour = deepcopy(tour1)
                if newCost <
                        status.bestCostToDate {
                    status.bestCostToDate =
                                        newCost
                    status.bestTour =
                                deepcopy(tour1)
```

```
                    fmt.Printf("\nLowest cost
                    tour to-date = %0.2f at
                    Temperature = %0.2f  Best
                    tour: %v",
                        status.bestCostToDate,
                        status.temperature,
                        status.bestTour)
                }
            } else {
                metropolis :=
                        math.Exp((status.previousCost
                        - newCost) /
                            status.temperature)
                r := rand.Float64()
                if r <= metropolis {// Uphill move
                    status.uphillMoves += 1
                    status.previousCost = newCost
                    status.tour = deepcopy(tour1)
                    if newCost <
                        status.bestCostToDate {
                        status.bestCostToDate =
                            newCost
                        status.bestTour =
                            deepcopy(tour1)
                        fmt.Printf("\nLowest cost
                        tour to-date = %0.2f at
                        Temperature = %0.2f  Best
                        tour: %v",
                        status.bestCostToDate,
                        status.temperature,
                        status.bestTour)
                    } else {
                        status.rejectedMoves += 1
                    }
                }
            }
```

```go
} else if newCost == cost2 {
    status.swapOps += 1
    // Determine whether to accept this
    // tour2
    if newCost < status.previousCost {
        status.downhillMoves += 1
        status.previousCost = newCost
        status.tour = deepcopy(tour2)
        if newCost <
                status.bestCostToDate {
            status.bestCostToDate =
                    newCost
            status.bestTour =
                    deepcopy(tour2)
            fmt.Printf("\nLowest cost
            tour to-date = %0.2f at
            Temperature = %0.2f  Best
           tour: %v", status.bestCostToDate,
           status.temperature,
           status.bestTour)
        }
    } else {
        metropolis :=
                math.Exp((status.previousCost
                - newCost) /
                    status.temperature)
        r := rand.Float64()
        if r <= metropolis {// Uphill move
            status.uphillMoves += 1
            status.previousCost = newCost
            status.tour = deepcopy(tour2)
            if newCost <
                status.bestCostToDate {
                status.bestCostToDate =
                    newCost
```

```go
                    status.bestTour =
                            deepcopy(tour2)
                    fmt.Printf("\nLowest cost
                    tour to-date = %0.2f at
                    Temperature = %0.2f  Best
                    tour: %v",
                    status.bestCostToDate,
                    status.temperature,
                    status.bestTour)
                } else {
                    status.rejectedMoves += 1
                }
            }
        }
    } else if newCost == cost3 {
        status.insertOps += 1
        // Determine whether to accept this
        // tour3
        if newCost < status.previousCost {
            status.downhillMoves += 1
            status.previousCost = newCost
            status.tour = deepcopy(tour3)
            if newCost <
                status.bestCostToDate {
                status.bestCostToDate = newCost
                status.bestTour =
                        deepcopy(tour3)
                fmt.Printf("\nLowest cost tour
                to-date = %0.2f at Temperature
                = %0.2f  Best tour: %v",
                status.bestCostToDate,
                status.temperature,
                status.bestTour)
            }
```

```go
                } else {
                    metropolis :=
                        math.Exp((status.previousCost
                        - newCost) /
                        status.temperature)
                r := rand.Float64()
                if r <= metropolis {// Uphill move
                    status.uphillMoves += 1
                    status.previousCost = newCost
                    status.tour = deepcopy(tour3)
                    if newCost <
                        status.bestCostToDate {
                        status.bestCostToDate =
                            newCost
                        status.bestTour =
                            deepcopy(tour3)
                        fmt.Printf("\nLowest cost
                        tour to-date = %0.2f at
                        Temperature = %0.2f  Best
                        tour: %v",
                        status.bestCostToDate,
                        status.temperature,
                        status.bestTour)
                    } else {
                        status.rejectedMoves += 1
                    }
                }
            }
        }
    }
}
// Cooling curve
if status.temperature >= 1000.0 {
    status.temperature *= 0.90
} else if status.temperature >= 500 {
    status.temperature *= 0.94
```

```
        } else if status.temperature >= 200 {
            status.temperature *= 0.97
        } else if status.temperature >= 50 {
            status.temperature *= 0.98
        } else {
            status.temperature *= 0.99
        }
    }
}

func main() {
    cities := []Point{}
    pt1 := Point{1150.0,1760.0}
    cities = append(cities, pt1)
    pt2 := Point{630.0, 1660.0}
    cities = append(cities, pt2)
    pt3 := Point{40.0, 2090.0}
    cities = append(cities, pt3)
    pt4 := Point{750.0, 1100.0}
    cities = append(cities, pt4)
    pt5 := Point{750.0, 2030.0}
    cities = append(cities, pt5)
    pt6 := Point{1030.0, 2070.0}
    cities = append(cities, pt6)
    pt7 := Point{1650.0, 650.0}
    cities = append(cities, pt7)
    pt8 := Point{1490.0, 1630.0}
    cities = append(cities, pt8)
    pt9 := Point{790.0, 2260.0}
    cities = append(cities, pt9)
    pt10 := Point{710.0, 1310.0}
    cities = append(cities, pt10)
    pt11 := Point{840.0, 550.0}
    cities = append(cities, pt11)
    pt12 := Point{1170.0, 2300.0}
    cities = append(cities, pt12)
```

```
pt13 := Point{970.0, 1340.0}
cities = append(cities, pt13)
pt14 := Point{510.0, 700.0}
cities = append(cities, pt14)
pt15 := Point{750.0, 900.0}
cities = append(cities, pt15)
pt16 := Point{1280.0, 1200.0}
cities = append(cities, pt16)
pt17 := Point{230.0, 590.0}
cities = append(cities, pt17)
pt18 := Point{460.0, 860.0}
cities = append(cities, pt18)
pt19 := Point{1040.0, 950.0}
cities = append(cities, pt19)
pt20 := Point{590.0, 1390.0}
cities = append(cities, pt20)
pt21 := Point{830.0, 1770.0}
cities = append(cities, pt21)
pt22 := Point{490.0, 500.0}
cities = append(cities, pt22)
pt23 := Point{1840.0, 1240.0}
cities = append(cities, pt23)
pt24 := Point{1260.0, 1500.0}
cities = append(cities, pt24)
pt25 := Point{1280.0, 790.0}
cities = append(cities, pt25)
pt26 := Point{490.0, 2130.0}
cities = append(cities, pt26)
pt27 := Point{1460.0, 1420.0}
cities = append(cities, pt27)
pt28 := Point{1260.0, 1910.0}
cities = append(cities, pt28)
pt29 := Point{360.0, 1980.0}
cities = append(cities, pt29)

graph := make([][]float64, NUMCITIES)
```

```
    for i:=0; i < NUMCITIES ; i++ {
        graph[i] = make([]float64, NUMCITIES)
    }
    createGraph(NUMCITIES, cities, graph)
    status.temperature = 2000.0
    simulatedAnnealing(graph)
    fmt.Printf("\nInverse Operations: %d  Swap Operations: %d  Insert
Operations: %d  Downhill moves: %d  Uphill moves, %d", status.inverseOps,
status.swapOps, status.insertOps, status.downhillMoves, status.uphillMoves)
}
/* Output
Cost of initial tour [0 1 2 3 4 5 6 7 8 9 10 11 12 13 14 15 16 17 18 19 20
21 22 23 24 25 26 27 28 0] is 25814.877363

Lowest cost tour to-date = 25669.20 at Temperature = 2000.00  Best tour:
[0 1 2 3 26 4 5 6 7 8 9 10 11 12 13 14 15 16 17 18 19 20 21 22 23 24 25
27 28 0]
Lowest cost tour to-date = 25456.00 at Temperature = 2000.00  Best tour:
[0 1 2 3 26 4 5 6 7 8 9 10 11 12 13 14 15 16 17 18 19 25 21 22 23 24 20
27 28 0]
Lowest cost tour to-date = 24872.68 at Temperature = 2000.00  Best tour:
[0 1 2 3 26 4 5 6 7 8 9 10 11 12 13 14 15 16 17 18 19 27 21 22 23 24 20
25 28 0]
Lowest cost tour to-date = 24249.12 at Temperature = 2000.00  Best tour:
[0 1 2 3 26 4 5 6 7 8 9 15 10 11 12 13 14 16 17 18 19 27 21 22 23 24 20
25 28 0]
Lowest cost tour to-date = 22921.49 at Temperature = 2000.00  Best tour:
[0 1 2 3 26 4 5 6 7 8 9 15 27 19 18 17 16 14 13 12 11 10 21 22 23 24 20
25 28 0]
Lowest cost tour to-date = 22479.54 at Temperature = 2000.00  Best tour:
[0 1 2 3 26 4 5 6 7 8 9 15 27 19 18 13 17 16 14 12 11 10 21 22 23 24 20
25 28 0]
Lowest cost tour to-date = 21640.15 at Temperature = 2000.00  Best tour:
[0 1 2 3 26 4 5 6 24 8 9 15 27 19 18 13 17 16 14 12 11 10 21 22 23 7 20
25 28 0]
```

Lowest cost tour to-date = 21208.12 at Temperature = 2000.00   Best tour:
[0 1 2 3 15 26 4 5 6 24 8 9 27 19 18 13 17 16 14 12 11 10 21 22 23 7 20
25 28 0]
Lowest cost tour to-date = 18984.25 at Temperature = 2000.00   Best tour:
[0 1 7 23 15 22 21 10 9 12 14 16 17 13 18 19 27 11 8 24 6 5 4 26 3 2 20
25 28 0]
Lowest cost tour to-date = 18849.81 at Temperature = 2000.00   Best tour:
[0 1 7 23 15 22 21 16 10 9 12 14 17 13 18 19 27 11 8 24 6 5 4 26 3 2 20
25 28 0]
Lowest cost tour to-date = 18735.36 at Temperature = 2000.00   Best tour:
[0 1 27 23 18 13 17 14 12 9 5 11 7 19 15 22 21 16 10 24 6 8 4 26 3 2 20
25 28 0]
Lowest cost tour to-date = 18218.92 at Temperature = 2000.00   Best tour:
[0 6 24 10 16 21 22 15 19 7 11 5 9 12 14 17 13 18 23 27 1 8 4 26 3 2 20
25 28 0]
Lowest cost tour to-date = 18110.89 at Temperature = 2000.00   Best tour:
[0 6 24 10 16 21 22 15 12 7 11 1 18 13 17 14 19 9 5 23 27 8 4 26 3 2 20
25 28 0]
Lowest cost tour to-date = 17873.48 at Temperature = 2000.00   Best tour:
[0 6 24 10 16 21 22 15 12 7 11 4 1 18 13 17 14 19 9 8 23 27 5 26 3 2 20
25 28 0]
Lowest cost tour to-date = 17544.27 at Temperature = 2000.00   Best tour:
[0 6 24 10 16 21 17 13 18 1 4 11 7 12 15 22 14 19 9 8 23 27 5 26 3 2 20
25 28 0]
Lowest cost tour to-date = 17327.28 at Temperature = 2000.00   Best tour:
[0 6 24 10 16 21 17 13 3 1 4 11 7 12 15 22 14 19 9 8 23 27 5 26 18 2 20
25 28 0]
Lowest cost tour to-date = 17300.34 at Temperature = 2000.00   Best tour:
[0 6 24 10 16 21 17 13 12 3 1 8 11 7 15 22 14 19 9 4 23 27 5 26 18 25 2
20 28 0]
Lowest cost tour to-date = 16585.81 at Temperature = 2000.00   Best tour:
[0 6 24 10 16 21 17 13 12 3 1 8 11 4 9 19 14 22 15 7 23 27 5 26 18 25 2
20 28 0]

Lowest cost tour to-date = 15655.98 at Temperature = 2000.00   Best tour:
[0 6 24 10 16 21 17 13 12 3 1 20 8 11 4 9 19 14 22 15 7 23 27 5 26 18
25 2 28 0]
Lowest cost tour to-date = 15655.04 at Temperature = 2000.00   Best tour:
[0 6 24 10 16 21 17 13 12 3 1 20 8 11 4 9 19 14 22 7 15 23 27 5 26 18
25 2 28 0]
Lowest cost tour to-date = 15439.96 at Temperature = 2000.00   Best tour:
[0 6 24 10 16 21 17 13 12 3 1 18 26 5 27 23 15 7 22 14 19 9 4 11 8 20
25 2 28 0]
Lowest cost tour to-date = 15315.97 at Temperature = 2000.00   Best tour:
[0 6 24 10 16 21 17 13 12 3 1 23 27 5 26 18 15 7 22 14 19 9 4 11 8 20
25 2 28 0]
Lowest cost tour to-date = 14339.52 at Temperature = 2000.00   Best tour:
[0 11 7 22 14 9 3 1 28 2 25 8 20 19 12 15 26 10 16 21 17 13 24 6 18 4 5
27 23 0]
Lowest cost tour to-date = 14118.61 at Temperature = 2000.00   Best tour:
[0 6 22 15 24 10 21 14 18 3 12 16 17 13 7 26 23 27 11 4 25 5 8 1 19 9 2
28 20 0]
Lowest cost tour to-date = 14082.51 at Temperature = 1800.00   Best tour:
[0 27 7 6 22 26 23 4 9 13 17 12 15 24 18 10 3 14 16 21 19 11 5 8 1 28 2
25 20 0]
Lowest cost tour to-date = 14009.54 at Temperature = 1800.00   Best tour:
[0 27 7 6 22 26 23 4 9 13 17 12 15 24 18 10 14 3 16 21 19 11 5 8 1 28 2
25 20 0]
Lowest cost tour to-date = 13848.14 at Temperature = 1800.00   Best tour:
[0 27 7 6 22 26 23 4 11 9 13 16 21 17 18 15 24 10 14 3 19 12 25 2 1 28
8 5 20 0]
Lowest cost tour to-date = 13659.70 at Temperature = 1800.00   Best tour:
[0 27 7 23 12 15 26 19 1 28 20 17 13 21 16 9 2 25 4 5 8 11 3 14 18 10
24 6 22 0]
Lowest cost tour to-date = 13386.10 at Temperature = 1800.00   Best tour:
[0 27 7 23 26 15 12 1 19 28 20 17 13 21 16 9 2 25 4 5 8 11 18 3 14 10
24 6 22 0]

Lowest cost tour to-date = 13074.00 at Temperature = 1620.00   Best tour:
[0 27 20 9 3 19 14 24 26 18 10 13 17 16 21 1 28 25 2 8 4 5 11 12 15 6
22 7 23 0]
Lowest cost tour to-date = 12666.24 at Temperature = 1180.98   Best tour:
[0 20 28 25 2 4 7 8 5 11 27 1 19 6 22 24 18 14 10 21 13 16 17 3 9 12 15
26 23 0]
Lowest cost tour to-date = 12643.30 at Temperature = 794.53   Best tour:
[0 27 5 11 8 20 25 2 28 4 19 1 9 7 26 22 24 6 14 10 13 16 21 17 3 23 15
12 18 0]
Lowest cost tour to-date = 12487.86 at Temperature = 746.86   Best tour:
[0 23 15 12 9 3 18 6 24 22 26 5 11 20 1 25 8 4 28 2 19 17 21 13 16 10
14 7 27 0]
Lowest cost tour to-date = 12162.28 at Temperature = 702.05   Best tour:
[0 20 5 11 4 25 2 28 8 9 19 3 17 16 21 13 10 14 15 12 1 18 6 24 26 22 7
23 27 0]
Lowest cost tour to-date = 11764.31 at Temperature = 702.05   Best tour:
[0 27 5 11 8 4 28 2 25 20 3 12 15 18 7 26 23 1 19 9 16 21 10 13 17 14
24 6 22 0]
Lowest cost tour to-date = 11640.94 at Temperature = 484.32   Best tour:
[0 7 15 19 14 3 12 20 8 4 5 23 26 22 6 24 18 10 21 13 16 17 9 1 28 2 25
11 27 0]
Lowest cost tour to-date = 11428.50 at Temperature = 484.32   Best tour:
[0 5 4 25 2 20 19 12 3 14 9 17 16 13 21 10 18 24 6 22 7 23 26 15 1 28 8
11 27 0]
Lowest cost tour to-date = 11411.98 at Temperature = 469.79   Best tour:
[0 27 23 26 7 22 15 19 3 9 14 13 17 16 21 10 6 24 18 12 20 1 11 5 4 25
28 2 8 0]
Lowest cost tour to-date = 11140.76 at Temperature = 469.79   Best tour:
[0 27 23 7 26 15 22 6 24 18 10 14 21 13 16 3 17 19 9 12 1 20 11 5 4 25
2 28 8 0]
Lowest cost tour to-date = 11091.14 at Temperature = 455.70   Best tour:
[0 27 11 5 8 4 28 2 25 20 1 12 24 6 10 21 16 17 13 18 9 19 3 14 15 23 7
22 26 0]

Lowest cost tour to-date = 10549.20 at Temperature = 346.44   Best tour:
[0 27 11 5 4 8 25 28 2 19 9 14 3 17 16 10 21 13 18 24 6 22 26 15 23 7
12 1 20 0]
Lowest cost tour to-date = 10477.04 at Temperature = 346.44   Best tour:
[0 5 27 11 4 8 25 28 2 19 9 14 3 17 21 16 13 10 18 24 6 22 26 15 7 23
12 1 20 0]
Lowest cost tour to-date = 10368.42 at Temperature = 325.96   Best tour:
[0 27 5 11 8 2 28 25 4 20 1 19 9 3 18 12 10 21 17 16 13 14 24 6 22 26 7
23 15 0]
Lowest cost tour to-date = 10162.14 at Temperature = 325.96   Best tour:
[0 27 11 5 8 2 28 25 4 20 12 1 19 9 3 14 10 21 17 16 13 18 24 6 22 26 7
23 15 0]
Lowest cost tour to-date = 9899.85 at Temperature = 247.81   Best tour: [0 5
27 11 8 25 2 28 4 20 1 19 9 12 15 18 14 13 3 17 16 21 10 24 6 22 26 7 23 0]
Lowest cost tour to-date = 9846.42 at Temperature = 226.17   Best tour: [0 7
23 26 15 22 6 24 18 3 13 21 16 17 10 14 12 9 19 1 20 28 2 25 4 8 5 11 27 0]
Lowest cost tour to-date = 9829.90 at Temperature = 179.14   Best tour:
[0 23 15 7 26 22 6 24 18 14 10 21 16 13 17 3 9 12 19 1 20 11 8 28 2 25
4 5 27 0]
Lowest cost tour to-date = 9740.09 at Temperature = 179.14   Best tour:
[0 23 7 26 15 22 6 24 18 14 10 21 16 13 17 3 9 12 19 1 20 11 8 28 2 25
4 5 27 0]
Lowest cost tour to-date = 9677.66 at Temperature = 175.56   Best tour:
[0 23 7 22 26 15 12 3 14 17 16 13 21 10 6 24 18 9 19 1 20 4 28 2 25 8
11 5 27 0]
Lowest cost tour to-date = 9642.94 at Temperature = 175.56   Best tour:
[0 23 7 26 22 15 12 3 14 17 16 13 21 10 6 24 18 9 19 1 20 4 28 2 25 8
11 5 27 0]
Lowest cost tour to-date = 9606.44 at Temperature = 175.56   Best tour:
[0 23 7 26 22 6 24 15 12 14 21 16 17 13 10 18 3 9 19 1 20 4 28 2 25 8
11 5 27 0]
Lowest cost tour to-date = 9596.98 at Temperature = 175.56   Best tour:
[0 23 7 26 22 6 24 10 13 17 16 21 14 12 15 18 3 9 19 1 20 4 28 2 25 8
11 5 27 0]

Lowest cost tour to-date = 9569.98 at Temperature = 175.56 Best tour:
[0 27 11 5 8 4 28 2 25 20 1 19 9 3 10 21 16 13 17 14 18 12 15 24 6 22 7
26 23 0]
Lowest cost tour to-date = 9490.31 at Temperature = 172.05 Best tour:
[0 27 20 5 11 8 4 25 2 28 1 19 9 3 12 15 18 14 17 16 21 13 10 24 6 22
26 7 23 0]
Lowest cost tour to-date = 9406.00 at Temperature = 161.93 Best tour:
[0 23 7 26 22 6 24 15 18 10 16 21 13 17 14 3 9 12 19 1 20 4 28 2 25 8
11 5 27 0]
Lowest cost tour to-date = 9248.08 at Temperature = 161.93 Best tour:
[0 23 7 26 22 6 24 15 18 10 21 16 13 17 14 3 9 12 19 1 20 4 28 2 25 8
11 5 27 0]
Lowest cost tour to-date = 9248.08 at Temperature = 161.93 Best tour:
[0 27 5 11 8 25 2 28 4 20 1 19 12 9 3 14 17 13 16 21 10 18 15 24 6 22
26 7 23 0]
Lowest cost tour to-date = 9120.82 at Temperature = 161.93 Best tour:
[0 27 5 11 8 25 2 28 4 20 1 19 9 12 3 14 17 13 16 21 10 18 15 24 6 22
26 7 23 0]
Lowest cost tour to-date = 9107.19 at Temperature = 93.85 Best tour: [0 27
5 11 8 25 2 28 4 20 1 19 9 12 3 14 17 13 16 21 10 18 24 6 22 26 15 23 7 0]
Lowest cost tour to-date = 9077.92 at Temperature = 93.85 Best tour: [0 27
5 11 8 25 2 28 4 20 1 19 9 12 3 14 17 13 16 21 10 18 24 6 22 15 26 7 23 0]
Lowest cost tour to-date = 9076.98 at Temperature = 72.17 Best tour: [0 23
15 26 7 22 6 24 18 10 21 16 13 17 14 3 12 9 19 1 20 4 28 2 25 8 11 5 27 0]
Lowest cost tour to-date = 9074.15 at Temperature = 72.17 Best tour: [0 27
5 11 8 25 2 28 4 20 1 19 9 3 14 17 13 16 21 10 18 24 6 22 7 26 15 12 23 0]
Inverse Operations: 748742 Swap Operations: 345285 Insert Operations:
625973 Downhill moves: 95259 Uphill moves, 96014
> Elapsed: 2.288s
*/

Discussion of Code

Let's focus on function ***simulatedAnnealing***.

A for-loop runs if temperature is greater than ***lowestTemperature***, which we set at 5.0. We consider perturbing the existing tour using the three operations discussed earlier. We assign the new tour to the tour among the three choices with the lowest tour cost.

If the cost of this new tour is smaller than the previous tour cost, we accept this new tour, increment the number of downhill moves, and update other status values.

Otherwise, we compute a ***metropolis*** value using the Boltzmann-like function to determine whether we accept an uphill move (a tour cost greater than the previous tour cost).

```
metropolis :=
   math.Exp((status.previousCost - newCost) /
               status.temperature)
```

Next, we generate a random float value between 0 and 1. If this value is less than ***metropolis***, we accept an uphill move by changing the tour to the tour with worse cost (uphill move) and then update the appropriate status values.

If the random float value is equal or greater than the ***metropolis*** value, we do not modify the current tour.

Following this, a new potential modification to the tour occurs using the three operations defined earlier. This continues until we have performed the requisite number of modifications specified for the given temperature. Then we use the logic of the cooling curve to lower the temperature and start the process just described again.

Results

The execution time of this run is 2.3 seconds on an iMac. The lowest-cost tour of **9074.15** is the known optimum tour for this 29-city problem. It is not unusual for the simulated annealing heuristic algorithm to find the optimum tour, although this is not guaranteed.

Displaying Final Results

If the code from Listing 17-3 is added to Listing 19-1 and the line ***DrawTour*** is added as the last line in function ***simulatedAnnealing***, we obtain the drawing shown in Figure 19-1.

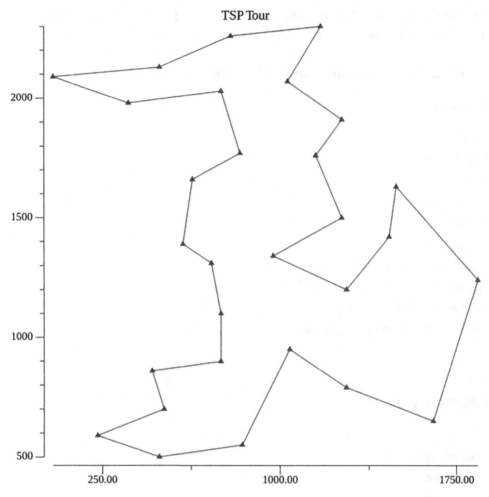

Figure 19-1. *A 29-city tour from simulated annealing*

Lines Crossing

As expected, this tour has no lines crossing. It is well known that if two edges in a closed polygon cross, there is a polygon with the same vertices that has a smaller perimeter. This follows from the triangle inequality. This inequality is that in any triangle, the sum of any two sides must be greater than the third side.

So **a necessary condition** for a tour to be optimum is that no lines cross in the tour. But that **is not a sufficient condition**. It is possible for suboptimal tours to not have lines cross.

The additional code for producing the graphical output is shown in Listing 19-2. Only the changed functions are shown.

Listing 19-2. Simulated annealing with graphical output

```go
package main

import (
    "fmt"
    "math"
    "math/rand"
    "time"
    "image/color"
    "gonum.org/v1/plot"
    "gonum.org/v1/plot/plotter"
    "gonum.org/v1/plot/vg"
    "gonum.org/v1/plot/vg/draw"
)

const (
    NUMCITIES = 29
)

type Point struct {
    x float64
    y float64
}

var cities []Point

func init() {
    // Snip
}

func (pt Point) distance(other Point) float64 {
    // Snip
}
```

```go
func createGraph(numCities int, cities []Point, graph
        [][]float64) {
    // Snip
}

func cost(graph [][]float64, tour []int) float64 {
    // Snip
}

func swap(tour []int) []int {
    // Snip
}

func insert(tour []int) []int {
    // snip
}

type Status struct {
    // Snip
}

var status Status

func deepcopy(tour []int) []int {
    // Snip
}

func simulatedAnnealing(graph [][]float64) {
    // Snip
    DrawTour(cities, status.bestTour)
}

func definePoints(cities []Point, tour []int)
    plotter.XYs {
    pts := make(plotter.XYs, len(cities) + 1)
    pts[0].X = cities[0].x
    pts[0].Y = cities[0].y
    for i := 1; i < len(cities); i++ {
```

```go
        pts[i].X = cities[tour[i]].x
        pts[i].Y = cities[tour[i]].y
    }
    pts[len(cities)].X = cities[0].x
    pts[len(cities)].Y = cities[0].y
    return pts
}

func DrawTour(cities []Point, tour []int) {
    data := definePoints(cities, tour) // plotter.XYs
    p := plot.New()
    p.Title.Text = "TSP Tour"
    lines, points, err := plotter.NewLinePoints(data)
    if err != nil {
        panic(err)
    }
    lines.Color = color.RGBA{R: 255, A: 255}
    points.Shape = draw.PyramidGlyph{}
    points.Color = color.RGBA{B: 255, A: 255}
    p.Add(lines, points)
    // Save the plot to a PNG file.
    if err := p.Save(6*vg.Inch, 6*vg.Inch, "tour.png");
            err != nil {
        panic(err)
    }
}

func main() {
    // Snip
}
```

19.4 Summary

This chapter presented a simulated annealing heuristic algorithm for solving TSP. The steps of this algorithm mimic the annealing of metal beams where the goal is to immerse the beam in a hot liquid and cool the beam slowly until the internal average energy of the lattice structure is minimized. Using an artificial temperature variable, the simulated annealing algorithm cools the solution space slowly while attempting to lower the cost of a tour.

We obtained remarkable results applying this heuristic algorithm to a 29-city problem.

The next chapter presents another heuristic algorithm for tackling TSP, a genetic algorithm.

## CHAPTER 20

Genetic Algorithm for TSP

The previous chapter presented an implementation of simulated annealing, a powerful and useful heuristic algorithm for solving TSP. We saw that this heuristic algorithm often obtains the optimum solution to the problem with relatively little computational effort.

This chapter presents another heuristic approach for TSP – genetic algorithm.

In the next section, we introduce the basis for this heuristic algorithm

20.1 Genetic Algorithm

A genetic algorithm is inspired by the biological maxim **survival of the fittest**.

As species evolve, traits that resonate with the ecosystem in which the species exist prevail. The traits that promote the greatest ability to survive in a hostile environment become dominant, and the traits that promote weakness disappear over time. This evolutionary process assumes continual changes in the underlying genetic structure of the species resulting from reproduction.

We apply this evolutionary model to TSP.

High-Level Description of Genetic Algorithm

An initial population of tours that visit each city once and return to the starting city is randomly generated. The fitness of each tour is the reciprocal of the tour cost. The smaller the tour cost, the higher the fitness.

We define a mating process as combining two tours to produce two offspring tours that are formed by some combination of the parent tours.

We define a mating pool as a collection of parent tours to be combined (mated) to produce the next generation of tours.

We define a mutation of a tour as a new tour that results from some small random perturbation of an existing tour.

© Richard Wiener 2022
R. Wiener, *Generic Data Structures and Algorithms in Go*, https://doi.org/10.1007/978-1-4842-8191-8_20

More Detailed Description of Genetic Algorithm

Initial Generation: We define a constant population size, say, **PopSize**. We generate **PopSize** random tours, each starting and ending at city 0. These tours represent our initial generation.

Rank the Population: We rank the tours that comprise the initial generation based on the fitness of each tour. The smaller the cost of the tour, the higher the fitness. We sort the tours by their fitness.

Mating Pool: We use a tournament selection rule that works as follows: A specified group of tours are randomly selected from the population, and the one with the highest fitness in the group is chosen as the first parent. This is repeated to choose the second parent. We continue this process until we have created **PopSize / 2** mating pairs.

Mating: This is the most challenging and important aspect of the genetic algorithm. We need to combine two tours to produce two child tours where each child retains a portion of its parents. We will utilize several crossover algorithms to accomplishing this mating.

Mutation: We will use a simple swap of two randomly chosen cities in a tour to produce a new tour. Such a new tour may have inferior fitness compared to the tour being mutated. This is like an uphill move in the simulated annealing algorithm. It promotes diversity and helps stave off a premature descent to a local minimum in the solution space. We apply mutation to a randomly selected small percentage of a given generation.

The steps for our genetic algorithm are the following:

1. Form an initial population of random tours of size **ToursPerGeneration**.

2. Select a small elite group of the fittest tours in the existing generation to be moved to the next generation.

3. Use tournament selection on the tours that remain to define a mating pool.

4. Perform mating of the parents in the mating pool to form a new generation of tours. This new generation contains the elite group from the previous generation along with the new children formed by mating.

5. Perform mutation on a randomly selected small fraction of the new generation.

6. Repeat steps 2 through 5 for a specified number of generations. Output the best tour to date as the tours progress from one generation to another.

In the next section, we construct our solution, following these steps.

20.2 Implementation of Genetic Algorithm
Step 1 – Form an Initial Population of Random Tours

We form an initial population of tours as follows:

```
var population [][]int

func CreateInitialPopulation() {
    firstCities := make([]int, NUMCITIES - 1)
    for i := 1; i < NUMCITIES; i++ {
        firstCities[i - 1] = i
    }
    for row := 0; row < ToursPerGeneration; row++ {
        rand.Shuffle(len(firstCities), func(i, j int) {
            firstCities[i], firstCities[j] =
                    firstCities[j], firstCities[i]
        })
        population[row] = []int{0}
        for col := 1; col < NUMCITIES; col++ {
            population[row] = append(population[row],
                        firstCities[col - 1])
        }
    }
}
```

We utilize the **Shuffle** function from package **rand** by producing a random sequence of integers from 1 to **NUMCITIES**. We initialize the global *population* variable at each row with value 0 and then append the random sequence to this initial value.

When we are done, each row of the population matrix contains a sequence of cities, each starting with city 0. When we compute the cost of each tour, we add the cost of going from the last city in the sequence back to city 0.

The **Cost** function for a given tour (row of the population matrix) is the following:

```go
func Cost(graph [][]float64, tour []int) float64 {
    result := 0.0
    for index := 0; index < len(tour) - 2; index++ {
        result += graph[tour[index]][tour[index+1]]
    }
    result += graph[tour[NUMCITIES - 1]][tour[0]]
    return result
}
```

Step 2 – Form an Elite Group of Best Tours

The function **ChooseEliteGroup**() returns a matrix of ELITENU best tours in the current population.

This function is given as follows:

```go
func ChooseEliteGroup() (elite [][]int) {
    // The population is sorted prior calling
    // this function

    // Initialize elite
    elite = make([][]int, EliteNumber)
    for row := 0; row < EliteNumber; row++ {
        elite[row] = make([]int, EliteNumber)
    }

    for row := 0; row < EliteNumber; row++ {
        elite[row] = DeepCopy(population[row])
    }
    return elite
}
```

The **DeepCopy** is needed because we wish to copy the values in each row of the sorted population and not the address of the row.

Step 3 – Tournament Selection

To obtain mating pairs from the current population minus the elite tours, we grab
TournamentNumber tours chosen randomly from the population minus elite tours, sort
them, and return the best tour (lowest cost) among them. That tour is parent1. We do
the same again to produce parent2. We mate the two parents and add the children into
newpopulation matrix.

We choose *EliteNumber* so that *ToursPerGeneration – EliteNumber* is an even
number. The number of parent pairs that we need to mate is (*ToursPerGeneration –
EliteNumber*) / 2.

Step 4 – Mating of Parents

The *OrderedCrossover* function which we use to mate two parent tours transmits
information about the relative ordering of the parents to the children.

We create two random crossover points in the parents and copy the segment
between them from parent1 to child1.

Starting from the second crossover point in parent2, we copy the remaining numbers
from the second parent to the first child, not allowing duplicates and wrapping when the
end of parent2 is encountered.

Do the same for the second child, reversing the role of the parents.

Consider the following example.

parent1: **0, 1, 2, | 3, 4, 5, 6, | 7, 8, 9**

parent2: **0, 8, 7, | 4, 3, 2, 1, | 9, 6, 5**

Here, the crossover indices are 3 and 6 shown with the vertical lines.

Let's walk through the process of obtaining child1.

After copying from parent1, child1 is the following:

x, x, x, 3, 4, 5, 6, x, x, x

Starting with the 9 in parent2 and working to the right and wrapping back to the
beginning of parent2 and child1, we add the values not in child1 to get

7, 2, 1, 3, 4, 5, 6, 9, 0, 8

Reversing the roles of parent1 and parent2, show that child2 is

0, 5, 6, 4, 3, 2, 1, 7, 8, 9

Now the challenge is to write function *OrderedCrossover* that implements the
preceding logic.

The logic is nontrivial. Function *OrderedCrossover* is the following:

```go
func OrderedCrossOver(parent1, parent2 []int) (child1,
         child2 []int) {
    var index1, index2 int
    n := len(parent1)
    for {
        index1 = 1 + rand.Intn(len(parent1)-1)
        index2 = 1 + rand.Intn(len(parent1)-1)
        if index1 != index2 {
            break // the two indices are different
        }
    }
    if index1 > index2 {
        index1, index2 = index2, index1
    }

    child1 = make([]int, len(parent1))
    child2 = make([]int, len(parent1))
    for i := 0; i < len(parent1); i++ {
        // Since 0 is a legal value
        child1[i] = -1
        child2[i] = -1
    }

    // Logic for child1
    for i := index1; i <= index2; i++ {
        child1[i] = parent1[i]
    }
    k := index2 + 1 // index for child1
    for i := index2 + 1; i < len(parent1); i++ {
        found, _ := In(parent2[i], child1)
        if !found {
            child1[k%n] = parent2[i]
            k += 1
        }
    }
```

```
for i := 0; i <= index2; i++ {
    found, _ := In(parent2[i], child1)
    if !found {
        // j := (i + index2 + 1) % n
        child1[k%n] = parent2[i]
        k += 1
    }
}

// Logic for child2
for i := index1; i <= index2; i++ {
    child2[i] = parent2[i]
}
k = index2 + 1 // index for child2
for i := index2 + 1; i < len(parent2); i++ {
    found, := In(parent1[i], child2)
    if !found {
        child2[k%n] = parent1[i]
        k += 1
    }
}
for i := 0; i <= index2; i++ {
    found, _ := In(parent1[i], child2)
    if !found {
        // j := (i + index2 + 1) % n
        child2[k%n] = parent1[i]
        k += 1
    }
}

// Form child11 and child22
// so they both start at 0
child11 := []int{}
child22 := []int{}
_, index0 := In(0, child1)
```

```
    for i := index0; i < len(child1); i++ {
        child11 = append(child11, child1[i])
    }
    for i := 0; i < index0; i++ {
        child11 = append(child11, child1[i])
    }

    _, index0 = In(0, child2)
    for i := index0; i < len(child2); i++ {
        child22 = append(child22, child2[i])
    }
    for i := 0; i < index0; i++ {
        child22 = append(child22, child2[i])
    }
    return child11, child22
}
```

We force each child to start their tour at city 0 by creating **child11** and **child22** from **child1** and **child2** in such a way that **child11** and **child22** start at city 0. The final portion of the **OrderedCrossover** function accomplishes this.

Helper function **In** is used in several places and is given as follows:

```
func In(value int, values []int) (bool, int) {
    // Returns true if value in values
    // returns index of location or -1 if not found
    for index := 0; index < len(values); index++ {
        if values[index] == value {
            return true, index
        }
    }
    return false, -1
}
```

Form Next Generation

We define a global variable ***newpopulation*** that is created from the global variable ***population***.

The new population consists of the elite tours from ***population***, the children from the parents that have been mated, and mutations that are performed with specified probability for each tour in the ***newpopulation***. These mutations involve swapping two randomly chosen cities in the tour.

The function for doing this is presented next.

```go
func FormNextGeneration() {
    elite := ChooseEliteGroup()
    // Move elite into newpopulation
    row := 0 // index into newpopulaton
    for ; row < EliteNumber; row++ {
        newpopulation[row] = DeepCopy(elite[row])
    }
    // Remove the first EliteNumber rows from
    // population
    population = population[EliteNumber:]

    // Initialize group1 and group2
    group1 := make([][]int, TournamentNumber)
    for i := 0; i < TournamentNumber; i++ {
        group1[i] = make([]int, NUMCITIES)
    }
    group2 := make([][]int, TournamentNumber)
    for i := 0; i < TournamentNumber; i++ {
        group2[i] = make([]int, NUMCITIES)
    }

    MatingPoolSize := (ToursPerGeneration -
                        EliteNumber) / 2
    for index := 0; index < MatingPoolSize; index++ {
        // Grap first group
        indicesChosen := []int{}
        rowsChosen := 0;
```

```go
for {
    randomRow := rand.Intn(TournamentNumber)
    found, _ := In(randomRow, indicesChosen)
    if !found {
        indicesChosen = append(indicesChosen,
            randomRow)
        group1[rowsChosen] =
            DeepCopy(population[randomRow])
        rowsChosen += 1
    }
    if rowsChosen == TournamentNumber {
        break
    }
}
// Grap second group
indicesChosen = []int{}
rowsChosen = 0;
for {
    randomRow := rand.Intn(TournamentNumber)
    found, _ := In(randomRow, indicesChosen)
    if !found {
        indicesChosen = append(indicesChosen,
            randomRow)
        group2[rowsChosen] =
            DeepCopy(population[randomRow])
        rowsChosen += 1
    }
    if rowsChosen == TournamentNumber {
        break
    }
}
// Sort group1 and group2
sort.Slice(group1, func(i, j int) bool {
    return Cost(group1[i]) < Cost(group1[j])
})
```

```
    sort.Slice(group2, func(i, j int) bool {
        return Cost(group2[i]) < Cost(group2[j])
    })
    parent1 := group1[0] // The best from group1
    parent2 := group2[0] // The best from group2
    child1, child2 := OrderedCrossOver(parent1,
                            parent2)
    newpopulation[row] = child1
    row += 1
    newpopulation[row] = child2
    row += 1
}
// Perform mutations
for row := 0; row < ToursPerGeneration; row++ {
    r := rand.Float64()
    if r <= ProbMutation {
        SwapMutation(newpopulation[row])
    }
}
population = make([][]int, ToursPerGeneration)
for i := 0; i < NUMCITIES; i++ {
    population[i] = make([]int, NUMCITIES)
}
// Copy newpopulation to population
for row := 0; row < ToursPerGeneration; row++ {
    for col := 0; col < NUMCITIES; col++ {
        population[row][col] =
                newpopulation[row][col]
    }
}
}
```

The code is heavily commented and should be straightforward to understand.

Sorting is accomplished using the **Slice** function from package **sort**.

```
sort.Slice(group1, func(i, j int) bool {
    return Cost(group1[i]) < Cost(group1[j])
})
```

Here, it is specified that the cost of a tour is the basis for sorting where lower-cost tours occur before higher-cost tours.

Putting the Pieces Together

In Listing 20-1, we present the entire program for solving the TSP with the heuristic genetic programming algorithm. We include a main driver that loads the same 29-city problem presented in Chapter 19 where it was tackled using simulated annealing. We present and compare the results of these two approaches to obtaining heuristic solutions to this TSP.

Listing 20-1. Genetic algorithm for TSP

```
package main

import (
    "fmt"
    "math"
    "math/rand"
    "sort"
    "time"
)

const (
    NUMCITIES        = 29
    EliteNumber      =  2
    ToursPerGeneration = 100
    NumberGenerations  = 50000
    TournamentNumber = 4
    ProbMutation     = 0.25
)
```

```go
type Point struct {
    x float64
    y float64
}

var population [][]int
var newpopulation [][]int
var graph [][]float64

func (pt Point) distance(other Point) float64 {
    dx := pt.x - other.x
    dy := pt.y - other.y
    return math.Sqrt(dx*dx + dy*dy)
}

func CreateGraph(numCities int, cities []Point,
                 graph [][]float64) {
    for row := 0; row < numCities; row++ {
        for col := 0; col < numCities; col++ {
            if row == col {
                graph[row][col] = 0.0
            } else {
                graph[row][col] =
                    cities[row].distance(cities[col])
            }
        }
    }
}

func DeepCopy(tour []int) []int {
    result := []int{}
    for i := range tour {
        result = append(result, tour[i])
    }
    return result
}
```

```go
func In(value int, values []int) (bool, int) {
    // Returns true if value in values
    // returns index of location or -1 if not found
    for index := 0; index < len(values); index++ {
        if values[index] == value {
            return true, index
        }
    }
    return false, -1
}

func Cost(tour []int) float64 {
    result := 0.0
    for index := 0; index < len(tour)-2; index++ {
        result += graph[tour[index]][tour[index+1]]
    }
    result += graph[tour[NUMCITIES-1]][tour[0]]
    return result
}

func CreateInitialPopulation() {
    firstCities := make([]int, NUMCITIES-1)
    for i := 1; i < NUMCITIES; i++ {
        firstCities[i-1] = i
    }
    for row := 0; row < ToursPerGeneration; row++ {
        rand.Shuffle(len(firstCities), func(i, j int) {
            firstCities[i], firstCities[j] =
                    firstCities[j], firstCities[i]
        })
        population[row] = []int{0}
        for col := 1; col < NUMCITIES; col++ {
            population[row] = append(population[row],
                                firstCities[col-1])
        }
    }
}
```

536

```go
func ChooseEliteGroup() (elite [][]int) {
    // The population is sorted prior calling
    // this function

    // Initialize elite
    elite = make([][]int, EliteNumber)
    for row := 0; row < EliteNumber; row++ {
        elite[row] = make([]int, EliteNumber)
    }

    for row := 0; row < EliteNumber; row++ {
        elite[row] = DeepCopy(population[row])
    }
    return elite
}

func FormNextGeneration() {
    elite := ChooseEliteGroup()
    // Move elite into newpopulation
    row := 0 // index into newpopulaton
    for ; row < EliteNumber; row++ {
        newpopulation[row] = DeepCopy(elite[row])
    }
    // Remove the first EliteNumber rows from
    // population
    population = population[EliteNumber:]

    // Initialize group1 and group2
    group1 := make([][]int, TournamentNumber)
    for i := 0; i < TournamentNumber; i++ {
        group1[i] = make([]int, NUMCITIES)
    }
    group2 := make([][]int, TournamentNumber)
    for i := 0; i < TournamentNumber; i++ {
        group2[i] = make([]int, NUMCITIES)
    }

    MatingPoolSize := (ToursPerGeneration -
                            EliteNumber) / 2
```

537

```
for index := 0; index < MatingPoolSize; index++ {
    // Grap first group
    indicesChosen := []int{}
    rowsChosen := 0;
    for {
        randomRow := rand.Intn(TournamentNumber)
        found, _ := In(randomRow, indicesChosen)
        if !found {
            indicesChosen = append(indicesChosen,
                               randomRow)
            group1[rowsChosen] =
                DeepCopy(population[randomRow])
            rowsChosen += 1
        }
        if rowsChosen == TournamentNumber {
            break
        }
    }
    // Grap second group
    indicesChosen = []int{}
    rowsChosen = 0;
    for {
        randomRow := rand.Intn(TournamentNumber)
        found, _ := In(randomRow, indicesChosen)
        if !found {
            indicesChosen = append(indicesChosen,
                               randomRow)
            group2[rowsChosen] =
                DeepCopy(population[randomRow])
            rowsChosen += 1
        }
        if rowsChosen == TournamentNumber {
            break
        }
    }
```

```go
    // Sort group1 and group2
    sort.Slice(group1, func(i, j int) bool {
        return Cost(group1[i]) < Cost(group1[j])
    })
    sort.Slice(group2, func(i, j int) bool {
        return Cost(group2[i]) < Cost(group2[j])
    })
    parent1 := group1[0] // The best from group1
    parent2 := group2[0] // The best from group2
    child1, child2 := OrderedCrossOver(parent1,
                                    parent2)
    newpopulation[row] = child1
    row += 1
    newpopulation[row] = child2
    row += 1
}
// Perform mutations
for row := 0; row < ToursPerGeneration; row++ {
    r := rand.Float64()
    if r <= ProbMutation {
        SwapMutation(newpopulation[row])
    }
}
population = make([][]int, ToursPerGeneration)
for i := 0; i < ToursPerGeneration; i++ {
    population[i] = make([]int, NUMCITIES)
}
// Copy newpopulation to population
for row := 0; row < ToursPerGeneration; row++ {
    for col := 0; col < NUMCITIES; col++ {
        population[row][col] =
                                newpopulation[row][col]
    }
}
}
```

```go
func SwapMutation(tour []int) {
    var index1, index2 int
    n := len(tour)
    for {
        index1 = 1 + rand.Intn(n-1)
        index2 = 1 + rand.Intn(n-1)
        if index2 != index1 + 4 {
            break // the two indices are different
        }
    }
    if index1 > index2 {
        index1, index2 = index2, index1
    }
    tour[index1], tour[index2] = tour[index2],
                                 tour[index1]

}

func OrderedCrossOver(parent1, parent2 []int)
            (child1, child2 []int) {
    var index1, index2 int
    n := len(parent1)
    for {
        index1 = 1 + rand.Intn(len(parent1)-1)
        index2 = 1 + rand.Intn(len(parent1)-1)
        if index1 != index2 {
            break // the two indices are different
        }
    }
    if index1 > index2 {
        index1, index2 = index2, index1
    }
    child1 = make([]int, len(parent1))
    child2 = make([]int, len(parent1))
    for i := 0; i < len(parent1); i++ {
        // Since 0 is a legal value
        child1[i] = -1
```

```
        child2[i] = -1
    }

    // Logic for child1
    for i := index1; i <= index2; i++ {
        child1[i] = parent1[i]
    }
    k := index2 + 1 // index for child1
    for i := index2 + 1; i < len(parent1); i++ {
        found, _ := In(parent2[i], child1)
        if !found {
            child1[k%n] = parent2[i]
            k += 1
        }
    }
    for i := 0; i <= index2; i++ {
        found, _ := In(parent2[i], child1)
        if !found {
            // j := (i + index2 + 1) % n
            child1[k%n] = parent2[i]
            k += 1
        }
    }

    // Logic for child2
    for i := index1; i <= index2; i++ {
        child2[i] = parent2[i]
    }
    k = index2 + 1 // index for child2
    for i := index2 + 1; i < len(parent2); i++ {
        found, _ := In(parent1[i], child2)
        if !found {
            child2[k%n] = parent1[i]
            k += 1
        }
    }
}
```

```go
    for i := 0; i <= index2; i++ {
        found, _ := In(parent1[i], child2)
        if !found {
            // j := (i + index2 + 1) % n
            child2[k%n] = parent1[i]
            k += 1
        }
    }
    // Form child11 and child22
    // so they both start at 0
    child11 := []int{}
    child22 := []int{}
    _, index0 := In(0, child1)
    for i := index0; i < len(child1); i++ {
        child11 = append(child11, child1[i])
    }
    for i := 0; i < index0; i++ {
        child11 = append(child11, child1[i])
    }

    _, index0 = In(0, child2)
    for i := index0; i < len(child2); i++ {
        child22 = append(child22, child2[i])
    }
    for i := 0; i < index0; i++ {
        child22 = append(child22, child2[i])
    }
    return child11, child22
}

func GeneticAlgorithm() {
    generation := 0
    population = make([][]int, ToursPerGeneration)
    for i := 0; i < ToursPerGeneration; i++ {
        population[i] = make([]int, NUMCITIES)
    }
```

```go
    newpopulation = make([][]int, ToursPerGeneration)
    for i := 0; i < ToursPerGeneration; i++ {
        newpopulation[i] = make([]int, NUMCITIES)
    }
    lowestCostTour := 1000000000.0
    CreateInitialPopulation()
    for {
        if generation == NumberGenerations {
            break
        }
        // Sort the population based on tour cost
        sort.Slice(population, func(i, j int) bool {
            return Cost(population[i]) <
                                    Cost(population[j])
        })
        bestCost := Cost(population[0])
        if bestCost < lowestCostTour {
            lowestCostTour = bestCost
            fmt.Printf("\nLowest cost tour at
            generation %d = %0.2f", generation,
            lowestCostTour)
        }
        FormNextGeneration()
        generation += 1
    }
}

func main() {
    rand.Seed(time.Now().UnixNano())
    cities := []Point{}
    // Known solution: 9074.15
    pt1 := Point{1150.0,1760.0}
    cities = append(cities, pt1)
    pt2 := Point{630.0, 1660.0}
    cities = append(cities, pt2)
    pt3 := Point{40.0, 2090.0}
```

```
cities = append(cities, pt3)
pt4 := Point{750.0, 1100.0}
cities = append(cities, pt4)
pt5 := Point{750.0, 2030.0}
cities = append(cities, pt5)
pt6 := Point{1030.0, 2070.0}
cities = append(cities, pt6)
pt7 := Point{1650.0, 650.0}
cities = append(cities, pt7)
pt8 := Point{1490.0, 1630.0}
cities = append(cities, pt8)
pt9 := Point{790.0, 2260.0}
cities = append(cities, pt9)
pt10 := Point{710.0, 1310.0}
cities = append(cities, pt10)
pt11 := Point{840.0, 550.0}
cities = append(cities, pt11)
pt12 := Point{1170.0, 2300.0}
cities = append(cities, pt12)
pt13 := Point{970.0, 1340.0}
cities = append(cities, pt13)
pt14 := Point{510.0, 700.0}
cities = append(cities, pt14)
pt15 := Point{750.0, 900.0}
cities = append(cities, pt15)
pt16 := Point{1280.0, 1200.0}
cities = append(cities, pt16)
pt17 := Point{230.0, 590.0}
cities = append(cities, pt17)
pt18 := Point{460.0, 860.0}
cities = append(cities, pt18)
pt19 := Point{1040.0, 950.0}
cities = append(cities, pt19)
pt20 := Point{590.0, 1390.0}
cities = append(cities, pt20)
```

```
    pt21 := Point{830.0, 1770.0}
    cities = append(cities, pt21)
    pt22 := Point{490.0, 500.0}
    cities = append(cities, pt22)
    pt23 := Point{1840.0, 1240.0}
    cities = append(cities, pt23)
    pt24 := Point{1260.0, 1500.0}
    cities = append(cities, pt24)
    pt25 := Point{1280.0, 790.0}
    cities = append(cities, pt25)
    pt26 := Point{490.0, 2130.0}
    cities = append(cities, pt26)
    pt27 := Point{1460.0, 1420.0}
    cities = append(cities, pt27)
    pt28 := Point{1260.0, 1910.0}
    cities = append(cities, pt28)
    pt29 := Point{360.0, 1980.0}
    cities = append(cities, pt29)

    graph = make([][]float64, NUMCITIES)
    for i:=0; i < NUMCITIES ; i++ {
        graph[i] = make([]float64, NUMCITIES)
    }
    CreateGraph(NUMCITIES, cities, graph)
    GeneticAlgorithm()
}

/* Output
Lowest cost tour at generation 0 = 22019.11
Lowest cost tour at generation 1 = 20169.35
Lowest cost tour at generation 5 = 20017.31
Lowest cost tour at generation 6 = 19545.05
Lowest cost tour at generation 7 = 18447.20
Lowest cost tour at generation 12 = 18340.78
Lowest cost tour at generation 13 = 17953.87
Lowest cost tour at generation 15 = 17350.68
```

```
Lowest cost tour at generation 16 = 17095.07
Lowest cost tour at generation 18 = 16612.21
Lowest cost tour at generation 19 = 16425.63
Lowest cost tour at generation 20 = 16299.86
Lowest cost tour at generation 24 = 16002.17
Lowest cost tour at generation 28 = 15749.40
Lowest cost tour at generation 30 = 14754.66
Lowest cost tour at generation 53 = 13900.84
Lowest cost tour at generation 68 = 13831.31
Lowest cost tour at generation 72 = 13668.22
Lowest cost tour at generation 73 = 13636.80
Lowest cost tour at generation 77 = 13392.64
Lowest cost tour at generation 92 = 12979.84
Lowest cost tour at generation 103 = 12200.31
Lowest cost tour at generation 123 = 12030.21
Lowest cost tour at generation 186 = 11960.10
Lowest cost tour at generation 191 = 11860.86
Lowest cost tour at generation 204 = 11647.36
Lowest cost tour at generation 209 = 11639.41
Lowest cost tour at generation 215 = 11582.62
Lowest cost tour at generation 218 = 11580.22
Lowest cost tour at generation 224 = 11255.27
Lowest cost tour at generation 280 = 11150.08
Lowest cost tour at generation 344 = 11099.42
Lowest cost tour at generation 423 = 10775.75
Lowest cost tour at generation 482 = 10717.58
Lowest cost tour at generation 492 = 10592.38
Lowest cost tour at generation 496 = 10587.10
Lowest cost tour at generation 503 = 10556.30
Lowest cost tour at generation 508 = 10489.54
Lowest cost tour at generation 513 = 10415.89
Lowest cost tour at generation 519 = 10409.44
Lowest cost tour at generation 527 = 10292.43
Lowest cost tour at generation 536 = 10256.38
Lowest cost tour at generation 561 = 9990.04
```

```
Lowest cost tour at generation 795 = 9936.06
Lowest cost tour at generation 810 = 9869.37
Lowest cost tour at generation 883 = 9817.69
Lowest cost tour at generation 891 = 9694.19
Lowest cost tour at generation 909 = 9616.14
Lowest cost tour at generation 956 = 9541.14
Lowest cost tour at generation 965 = 9456.43
Lowest cost tour at generation 970 = 9362.68
Lowest cost tour at generation 1179 = 9285.43
*/
```

It takes less than ten seconds for this program to terminate.

After ten runs, the lowest-cost tour at generation 1179 is **9285**. This is an error of 2 percent from the known optimum solution of **9074**.

Clearly, this approach to solving TSP is useful.

20.3 Summary

In this chapter, we presented an approach to solving TSP based on genetic modeling and survival of the fittest. As the program moves from one generation to another and solutions evolve, the best tours converge to approximate the optimum solution.

Each run of the genetic algorithm is a new experiment. The results are greatly dependent on the constants chosen.

In the next chapter, we turn our attention to machine learning and neural networks.

CHAPTER 21

Neural Networks and Machine Learning

The previous chapter presented an implementation of a generic algorithm for solving TSP.

This chapter introduces neural networks and machine learning. We present an implementation of a neural network from scratch.

In the next section, we present an overview of machine learning and neural networks.

21.1 Overview of Neural Networks and Machine Learning

AI (artificial intelligence) has its roots in research done at Dartmouth in 1956. Its goal is to mimic human reasoning.

Machine learning is a subfield of AI. It uses statistics, operations research, and neural network models to obtain insights from data. It allows the computer (machine) to obtain insights through an iterative process that mimics how we believe the human brain learns new things. It allows computers the ability to learn to perform complex tasks without explicitly being programmed.

Applications of machine learning include natural language processing, including language translation, image classification and analysis, chatbots, medical diagnosis, game playing, pattern recognition, and stock price prediction.

Machine learning starts with data, often a huge quantity of data. This data may be numerical time series, photos, text, repair records, bank transactions, sales reports, or sensor data from a multitude of sources (weather data, seismic data, medical data, etc.).

© Richard Wiener 2022
R. Wiener, *Generic Data Structures and Algorithms in Go*, https://doi.org/10.1007/978-1-4842-8191-8_21

Such data is used to "train" a computer model such as a neural network. After sufficient training, the model is used to perform classification, to predict future outcomes, make a move in a game, provide a translation of some text, etc., based on the learning achieved from the training data.

Training

Training a neural network model involves feeding forward input data through the network to one or more outputs, feeding back errors that are detected in the output to modify the network to minimize these errors. We will examine this important process in detail in this chapter.

Neural Networks

Neural networks are function approximators. By training such a network with many observed inputs and corresponding outputs, the goal is to obtain a reliable output when new inputs are applied.

For example, if our goal is to have our network distinguish between a photo of a dog and a photo of a cat, we train the network by sending in numerous cat and dog images, each time informing the network whether the image (a two-dimensional matrix of pixel values) is a dog or cat. Then when images the network has never seen are sent in, the network will hopefully determine with high accuracy whether it is a dog or cat image.

Such a trained neural network may be thought of as a mathematical function that when presented with input produces some output (a classification in this case).

A neural network generally contains

1. A collection of input values that comprise an input layer

2. One or more hidden layers

3. An output layer with output values

4. A collection of weights and biases between layers

5. An activation function for each layer

Perceptron

In 1943, neurophysiologist Warren McCulloch and mathematician Walter Pitts defined a simple model of a neuron that takes a set of inputs, multiplies them by weighted values and adds a bias value, and then puts them through an activation function, which produces an output of 0 or 1. This model is called the McCulloch-Pitts perceptron.

A schematic of this perceptron is shown in Figure 21-1.

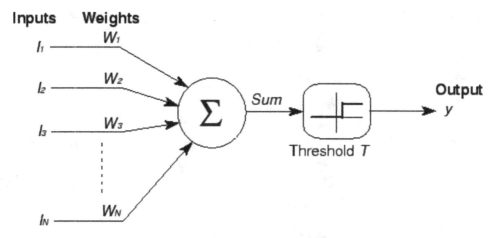

McCulloch-Pitts perceptron | Source: **Wikimedia Commons**

Figure 21-1. *Perceptron*

The output, y, is the sum of the input values multiplied by the set of weights, W, and a bias value b and followed by a threshold function that converts the sum to a value between zero and one.

Next, the expected value is compared to the actual value to form an error. This computed error is fed back to the weights to modify the weights to reduce this error. After many iterations, it is hoped that the error can be made very small.

The process of obtaining the output from the inputs and weights is called forward propagation. The process of modifying the weights based on the error is called backpropagation. By knowing the derivative of the error with respect to each weight, a recursive-descent algorithm that successively modifies the weights while decreasing the output error is achieved.

By stacking a collection of neurons in various layers, a neural network is created.

Schematics of Neural Networks

Schematics of two such neural networks are shown in Figures 21-2 and 21-3. The first has one "hidden" layer containing four neurons and a single output layer.

The second neural network has three layers containing four neurons each and is a deep neural network because of the many layers.

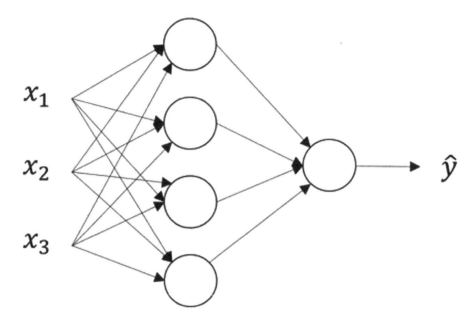

***Figure 21-2.** Neural network with one hidden layer*

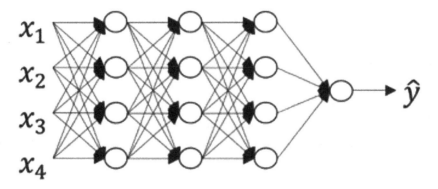

***Figure 21-3.** Neural network with many hidden layers*

A Neuron

We drill deeper and examine an individual neuron or node in a neural network. In Figure 21-4, we show such a neuron.

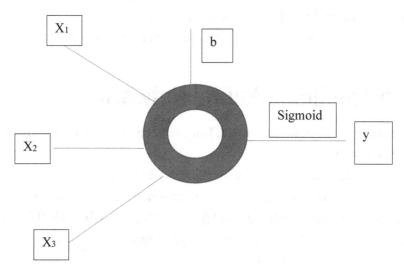

Figure 21-4. *A neuron*

The output, **y**, is computed from the inputs x_1, x_2, and x_3 as follows:

$$z = w_1x_1 + w_2x_2 + w_3x_3 + b$$

We follow this by taking this linear combination of inputs and bias and using a nonlinear activation function such as the sigmoid function as follows:

$$y = sigmoid(z) = 1 / (1 + e^{-z})$$

The use of the sigmoid activation function forces the result to be between 0 and 1 as z varies from a large negative number to a large positive number.

In the next section, we define a simple problem that we will solve.

21.2 A Concrete Example

Suppose we wish to construct and train a neural network to determine whether a diagnostic test indicates that a patient has a particular disease. There are two numbers from the test, **x** and **y**, each between 0.0 and 1.0. If $x^2 < y$, the test is negative; otherwise, it is positive. We will represent a negative test by the numerical label **0** and a positive test with the numerical label **1**.

The neural network will receive 150 test results (each result a pair of numbers, each between 0.0 and 1.0). The output of the neural network contains 150 computed scores, each between 0.0 and 1.0, as well as 150 correct labels, each between 0.0 and 1.0, based on the.

This example allows us to introduce the methodology of neural network modeling and computation and see how the important pieces fit together.

In the next section, we will build a neural network from scratch to solve this problem.

21.3 Constructing a Neural Network

We will construct a simple neural network, from scratch, that solves the problem presented in Section 21.2.

We define a weight matrix. Each column of this weight matrix shows the weights from all neurons in the previous layer to a particular neuron in the current layer.

So, for example, the values **w[0][2]**, **w[1][2]**, **w[2][[2]**, **..., w[n – 1][2]** (the third column of the weight matrix) represent the weights from the n neurons in the previous level to neuron 3 in the current level.

The neural network that we will build contains 150 nodes for the input layer (one node for each test result containing two numbers), 25 nodes for the hidden layer, and 150 nodes for the output layer.

Matrices That Represent Network

The input matrix is of dimension **150 × 2**. Each row of this matrix contains the test result numbers x and y.

The weight matrix that connects the input layer to the hidden layer is of dimension **2 × 25**.

The weight matrix that connects the hidden layer to the output layer is of dimension **25 × 1**.

The output of the neural network is of dimension **150 × 1**.

We will set the biases to zero for this example.

Some sample input and output would be

Input 1: **<0.42, 0.1>** (positive test since $0.42^2 > 0.1$)

Input 2: **<0.6, 0.8>** (negative test since $0.6^2 < 0.8$)

We will generate the test results by generating random x and y values for each test, each between 0.0 and 1.0.

We use the 150 test results to train the network. We then generate 25 more fresh test results randomly, as before. We then use the fully trained network to predict whether each of the 25 new tests is positive or negative and tabulate our errors.

In the next section, we present and explain the implementation of a neural network that classifies the results of the diagnostic tests.

21.4 Neural Network Implementation

We begin the implementation by defining some global variables and initializing all weights with random values from 0 to 1.

```go
package main

import (
    "fmt"
    "math"
    "math/rand"
    "time"
)

var (
    InputLayer  = 150
    HiddenLayer = 25
    OutputLayer = InputLayer
    Inputs      = 2 // x and y values
)

var weights1 [][]float64
var derivatives1 [][]float64
var weights2 [][]float64
var derivatives2 [][]float64
var input [][]float64

func InitializeWeights1() {
    weights1 = make([][]float64, Inputs)
    derivatives1 = make([][]float64, Inputs)
    for row := 0; row < Inputs; row++ {
        weights1[row] = make([]float64, HiddenLayer)
```

```
        derivatives1[row] = make([]float64,
                             HiddenLayer)
    }
    for row := 0; row < Inputs; row++ {
        for col := 0; col < HiddenLayer; col++ {
            weights1[row][col] = rand.Float64()
        }
    }
}

func InitializeWeights2() {
    weights2 = make([][]float64, HiddenLayer)
    derivatives2 = make([][]float64, HiddenLayer)
    for row := 0; row < HiddenLayer; row++ {
        weights2[row] = make([]float64, 1)
        derivatives2[row] = make([]float64,
                                  OutputLayer)
    }
    for row := 0; row < HiddenLayer; row++ {
        for col := 0; col < 1; col++ {
            weights2[row][col] = rand.Float64()
        }
    }
}
```

The **derivatives1** and **derivatives2** matrices will be explained later.

Next, we look at two functions: **trueOutput** and **cost**.

The **trueOutput** function evaluates **a[0]**, representing x, and **a[1]**, representing y, and returns 0 for a negative test result and 1 for a positive test result.

The **cost** function compares the values in column zero (the only column) of the neural network **output** with the correct values, squares each error, adds the errors, and divides by the number of errors.

```
func trueOutput(a []float64) float64 {
    if a[0] * a[0] <= a[1] {
        return 0.0
    } else {
```

```go
        return 1.0
    }
}

func cost(output [][]float64) float64 {
    result := 0.0
    for i := 0; i < InputLayer; i++ {
        correctAnswer := trueOutput(input[i])
        result += (output[i][0] - correctAnswer) * (output[i][0] -
        correctAnswer)
    }
    return result / float64(InputLayer)
}
```

The functions *dot*, *DotProduct*, and *Sigmoid* are support functions that support the neural network matrix operations that are needed.

```go
func dot(vector1 []float64, vector2 []float64) float64 {
    if len(vector1) != len(vector2) {
        panic("Illegal vector dimensions for dot product.")
    }
    result := 0.0
    for i := 0; i < len(vector1); i++ {
        result += vector1[i] * vector2[i]
    }
    return result
}

func DotProduct(matrix1, matrix2 [][]float64) (result [][]float64) {
    rows1 := len(matrix1)
    cols1 := len(matrix1[0])
    rows2 := len(matrix2)
    cols2 := len(matrix2[0])

    if cols1 != rows2 {
        panic("Cannot take dot product")
    }
    result = make([][]float64, rows1)
```

```go
    for row := 0; row < rows1; row++ {
        result[row] = make([]float64, cols2)
    }
    for row := 0; row < rows1; row++ {
        for col := 0; col < cols2; col++ {
            column := []float64{}
            for r := 0; r < cols1; r++ {
                column = append(column,
                            matrix2[r][col])
            }
            result[row][col] = dot(matrix1[row],
                                column)
        }
    }
    return result
}

func Sigmoid(matrix [][]float64) (result [][]float64) {
    rows := len(matrix)
    cols := len(matrix[0])
    result = make([][]float64, rows)
    for row := 0; row < rows; row++ {
        result[row] = make([]float64, cols)
    }
    for row := 0; row < rows; row++ {
        for col := 0; col < cols; col++ {
            result[row][col] = 1.0 / (1.0 + math.Exp(-
                            matrix[row][col]))
        }
    }
    return result
}
```

These functions are needed in the ***FeedForward*** function that transforms the neural network input to its output.

```
func FeedForward() [][]float64 {
    hidden := Sigmoid(DotProduct(input, weights1))
    output := Sigmoid(DotProduct(hidden, weights2))
    return output
}
```

We see that by taking the dot product of the ***input*** matrix with the ***weights1*** matrix and following this by the dot product of the resulting ***hidden*** matrix with the ***weights2*** matrix, we get the output matrix.

The ***Sigmoid*** activation function ensures that all the values are scaled to be between 0 and 1.

So far, we have examined how the inputs to the network (a matrix of 150 test results, each test having two real numbers) produce 150 outputs.

The magic of neural networks is the process of training the network. This means making modifications to the weight's matrices (***weights1*** and ***weights2*** in this case) and to lower the mean-squared average error between the computed output and the correct results (the **cost**).

The process for achieving this is called **backpropagation**.

The mathematics related to **backpropagation** is complex. See, for example, https://hmkcode.com/ai/backpropagation-step-by-step/.

Backpropagation involves taking partial derivatives of the cost with respect to each of the many weights. Each of these partial derivatives characterizes how the **cost** would be increased or decreased if a small change in a particular weight were made. If we knew the partial derivative for each of the weights, we could modify each weight with the goal of lowering the mean-squared error (the cost). The partial derivative would specify the direction and magnitude of the needed weight modification.

Since this chapter aims at introducing the mechanics of neural networks, we will bypass the mathematics by estimating the partial derivatives empirically. The price we pay for this is performance. At each iteration of backpropagation, we need to evaluate the effect of changing each weight on the overall cost of the network output.

Estimating the Partial Derivatives of Cost with Respect to Each Weight

For each weight in *weights1* and *weights2*, we add 0.01 or any other small amount to the weight. We compute the output of the network and its cost after making this change. The partial derivative of cost with respect to this weight is the ratio of the change in cost resulting from the tweak in the weight to the change in weight. If this ratio is positive, we save this positive partial derivative in a separate matrix with the same dimensions as the weight matrix. If the ratio is negative, we change the sign of the ratio and save it in the separate partial derivative matrix.

After we have estimated and saved all the partial derivatives, we modify the entire *weights1* and *weights2* matrices by the partial derivative amounts. This represents the first iteration of training in the network.

The function *ComputeDerivatives*, shown in the following, performs the estimation of partial derivatives:

```
func ComputeDerivatives() {
    // Estimates the partial derivative of the cost
    // with respect to each weight
    for row := 0; row < Inputs; row++ {
        for col := 0; col < HiddenLayer; col++ {
            output1 := FeedForward()
            c1 := cost(output1)
            weights1[row][col] += .01
            output2 := FeedForward()
            c2 := cost(output2)
            weights1[row][col] -= .01
            derivatives1[row][col] = (c2 - c1) / .01
            weights2[col][0] += .01
            output3 := FeedForward()
            c3 := cost(output3)
            weights2[col][0] -= .01
            derivatives2[col][0] = (c3 - c1) / .01
        }
    }
}
```

The function **BackPropagate** changes each weight, as shown, based on the values in the derivatives matrix.

```
func BackPropagate() {
    ComputeDerivatives()
    // Modifiy weights1 and weights2
    for row := 0; row < Inputs; row++ {
        for col := 0; col < HiddenLayer; col++ {
            weights1[row][col] -=
                    derivatives1[row][col]
        }
    }
    for row := 0; row < HiddenLayer; row++ {
        for col := 0; col < 1; col++ {
            weights2[row][col] -=
                    derivatives2[row][col]
        }
    }
}
```

Finally, the function **Train()** iteratively modifies the weights with the goal of lowering the cost.

```
func Train() {
    for epoch := 1; epoch < 1500; epoch++ {
        output := FeedForward()
        fmt.Println("cost = ", cost(output))
        BackPropagate()
    }
}
```

We put all the pieces together in Listing 21-1 including a main driver function that builds inputs for the neural network, trains the network, and outputs the results on fresh data after the training is completed.

Listing 21-1. Neural network from scratch

```go
package main

import (
    "fmt"
    "math"
    "math/rand"
    "time"
)

var (
    InputLayer  = 150
    HiddenLayer = 25
    OutputLayer = InputLayer
    Inputs      = 2 // x and y values
)

var weights1 [][]float64
var derivatives1 [][]float64
var weights2 [][]float64
var derivatives2 [][]float64
var input [][]float64

func InitializeWeights1() {
    weights1 = make([][]float64, Inputs)
    derivatives1 = make([][]float64, Inputs)
    for row := 0; row < Inputs; row++ {
        weights1[row] = make([]float64, HiddenLayer)
        derivatives1[row] = make([]float64,
                                   HiddenLayer)
    }
    for row := 0; row < Inputs; row++ {
        for col := 0; col < HiddenLayer; col++ {
            weights1[row][col] = rand.Float64()
        }
    }
}
```

```go
func InitializeWeights2() {
    weights2 = make([][]float64, HiddenLayer)
    derivatives2 = make([][]float64, HiddenLayer)
    for row := 0; row < HiddenLayer; row++ {
        weights2[row] = make([]float64, 1)
        derivatives2[row] = make([]float64,
                                 OutputLayer)
    }
    for row := 0; row < HiddenLayer; row++ {
        for col := 0; col < 1; col++ {
            weights2[row][col] = rand.Float64()
        }
    }
}


func dot(vector1 []float64,vector2 []float64) float64 {
    if len(vector1) != len(vector2) {
        panic("Illegal vector dimensions for dot
                     product.")
    }
    result := 0.0
    for i := 0; i < len(vector1); i++ {
        result += vector1[i] * vector2[i]
    }
    return result
}

func DotProduct(matrix1, matrix2 [][]float64) (result
                              [][]float64) {
    rows1 := len(matrix1)
    cols1 := len(matrix1[0])
    rows2 := len(matrix2)
    cols2 := len(matrix2[0])

    if cols1 != rows2 {
        panic("Cannot take dot product")
    }
```

```go
    result = make([][]float64, rows1)
    for row := 0; row < rows1; row++ {
        result[row] = make([]float64, cols2)
    }
    for row := 0; row < rows1; row++ {
        for col := 0; col < cols2; col++ {
            column := []float64{}
            for r := 0; r < cols1; r++ {
                column = append(column,
                                    matrix2[r][col])
            }
            result[row][col] = dot(matrix1[row],
                                    column)
        }
    }
    return result
}

func Sigmoid(matrix [][]float64) (result [][]float64) {
    rows := len(matrix)
    cols := len(matrix[0])
    result = make([][]float64, rows)
    for row := 0; row < rows; row++ {
        result[row] = make([]float64, cols)
    }
    for row := 0; row < rows; row++ {
        for col := 0; col < cols; col++ {
            result[row][col] = 1.0 / (1.0 + math.Exp(-
                    matrix[row][col]))
        }
    }
    return result
}

func trueOutput(a []float64) float64 {
    if a[0] * a[0] <= a[1] {
        return 0.0
```

```go
    } else {
        return 1.0
    }
}

func cost(output [][]float64) float64 {
    result := 0.0
    for i := 0; i < InputLayer; i++ {
        correctAnswer := trueOutput(input[i])
        result += (output[i][0] - correctAnswer) *
                        (output[i][0] - correctAnswer)
    }
    return result / float64(InputLayer)
}

func ComputeDerivatives() {
    // Estimates the partial derivative of the cost
    // with respect to each weight
    for row := 0; row < Inputs; row++ {
        for col := 0; col < HiddenLayer; col++ {
            output1 := FeedForward()
            c1 := cost(output1)
            weights1[row][col] += .01
            output2 := FeedForward()
            c2 := cost(output2)
            weights1[row][col] -= .01

            derivatives1[row][col] = (c2 - c1) / .01
            weights2[col][0] += .01
            output3 := FeedForward()
            c3 := cost(output3)
            weights2[col][0] -= .01
            derivatives2[col][0] = (c3 - c1) / .01
        }
    }
}
```

```go
func FeedForward() [][]float64 {
    hidden := Sigmoid(DotProduct(input, weights1))
    output := Sigmoid(DotProduct(hidden, weights2))
    return output
}

func BackPropagate() {
    ComputeDerivatives()
    // Modifiy weights1 and weights2
    for row := 0; row < Inputs; row++ {
        for col := 0; col < HiddenLayer; col++ {
            weights1[row][col] -=
                derivatives1[row][col]
        }
    }
    for row := 0; row < HiddenLayer; row++ {
        for col := 0; col < 1; col++ {
            weights2[row][col] -=
                derivatives2[row][col]
        }
    }
}

func Train() {
    for epoch := 1; epoch < 1500; epoch++ {
        output := FeedForward()
        fmt.Println("cost = ", cost(output))
        BackPropagate()
    }
}

func main() {
    rand.Seed(time.Now().UnixNano())
    InitializeWeights1()
    InitializeWeights2()
    input = make([][]float64, InputLayer)
```

```go
for row := 0; row < InputLayer; row++ {
    input[row] = make([]float64, Inputs)
}
for row := 0; row < InputLayer; row++ {
    for col := 0; col < Inputs; col++ {
        input[row][col] = rand.Float64()
        input[row][col] = rand.Float64()
    }
}

Train()
// Use existing weights and see how well
// the neural network handles new data
InputLayer = 25
OutputLayer = 25
input = make([][]float64, InputLayer)
for row := 0; row < InputLayer; row++ {
    input[row] = make([]float64, Inputs)
}
for row := 0; row < InputLayer; row++ {
    for col := 0; col < Inputs; col++ {
        input[row][col] = rand.Float64()
        input[row][col] = rand.Float64()
    }
}
output := FeedForward()
var verdict bool // false by default
for i := 0; i < InputLayer; i++ {
    if output[i][0] > 0.5 && trueOutput(input[i])
            == 1 {
        verdict = true
    } else if output[i][0] < 0.5 &&
            trueOutput(input[i]) == 0 {
        verdict = true
    }
```

```
        fmt.Printf("\nComputed value: %f  correct
        answer = %f  Correct Estimate: %v",
        output[i][0], trueOutput(input[i]), verdict)
    }
    fmt.Println()
}
```

In the next section, we examine the program output.

21.5 Output from Neural Network

The output is voluminous. The function ***Train*** performs 1500 epochs, each epoch involving a forward and back computation that trains the network. And each iteration causes an output of the current mean-squared error, cost. It is interesting and important to observe the evolution of these costs and see how they decrease as the network gets trained.

Only a portion of the output is shown in the interest of space. Of notice are the results of testing 25 fresh inputs. The outputs for these fresh inputs indicate 100 percent accuracy by the neural network if we interpret an output greater than 0.5 as positive and less than 0.5 as negative.

```
cost =  0.6637402659161185
cost =  0.6636630956218156
cost =  0.6635817634090139
cost =  0.6634959251095335
cost =  0.6634051977180305
cost =  0.6633091537786991
cost =  0.6632073147732753
cost =  0.663099143297417
cost =  0.6629840337592922
cost =  0.6628613012653831
cost =  0.6627301682691753
cost =  0.6625897484413662
cost =  0.6624390270658043
cost =  0.6622768370596327
cost =  0.6621018294396175
```

```
cost = 0.6619124366814639
cost = 0.6617068269041807
cost = 0.6614828460975539
cost = 0.6612379446084452
cost = 0.6609690826757143
cost = 0.660672607747008
cost = 0.660344093298625
cost = 0.6599781243955765
...
cost = 0.5811888364173224
cost = 0.5117698491474624
cost = 0.3431174223399081
cost = 0.22278592696925442
cost = 0.22219689682890306
cost = 0.22164595119640423
cost = 0.22109565104100137
cost = 0.22054536159623406
cost = 0.21999484909425368
cost = 0.21944389274059498
cost = 0.21889227410423057
cost = 0.2183397768384557
cost = 0.21778618670317892
cost = 0.21723129159816149
cost = 0.21667488160010032
cost = 0.21611674900371572
cost = 0.21555668836712247
cost = 0.21499449656177205
cost = 0.21442997282722046
cost = 0.21386291883096756
cost = 0.21329313873359648
cost = 0.21272043925942305
cost = 0.21214462977284917
cost = 0.21156552236059478
cost = 0.21098293191996614
cost = 0.2103966762532983
```

```
cost =   0.209806576168689
...
cost =   0.14066304727899404
cost =   0.1397261098810677
cost =   0.138792748311244
cost =   0.13786318684860793
cost =   0.13693764327438454
cost =   0.1360163286396859
cost =   0.13509944705792176
cost =   0.1341871955218915
cost =   0.13327976374542316
cost =   0.13237733402931087
cost =   0.13148008115118084
cost =   0.13058817227880457
cost =   0.12970176690062738
cost =   0.12882101681236946
cost =   0.1279460660403555
cost =   0.12707705089837232
cost =   0.1262140999795274
cost =   0.1253573342007353
cost =   0.12450686685930376
cost =   0.12366280370623699
cost =   0.12282524303519099
cost =   0.121994275786003
cost =   0.12116998566170542
cost =   0.12035244925792699
cost =   0.1195417362035959
cost =   0.11873790931186408
cost =   0.117941024740192
cost =   0.1171511321585531
cost =   0.11636827492474516
cost =   0.11559249026582975
cost =   0.11482380946475092
cost =   0.11406225805122268
...
```

```
cost =    0.016816878951632613
cost =    0.01680969687113106
cost =    0.016802523693735683
cost =    0.0167953594016083
cost =    0.01678820397696063
cost =    0.01678105740205403
cost =    0.016773919659199398
cost =    0.016766790730756983
cost =    0.016759670599136158
cost =    0.016752559246795311
cost =    0.016745456656241651
cost =    0.016738362810030993
cost =    0.016731277690767675
cost =    0.016724201281104255
```

Computed value: 0.991841   correct answer = 1.000000   Correct Estimate: true
Computed value: 0.001271   correct answer = 0.000000   Correct Estimate: true
Computed value: 0.025702   correct answer = 0.000000   Correct Estimate: true
Computed value: 0.998475   correct answer = 1.000000   Correct Estimate: true
Computed value: 0.958112   correct answer = 1.000000   Correct Estimate: true
Computed value: 0.021453   correct answer = 0.000000   Correct Estimate: true
Computed value: 0.000903   correct answer = 0.000000   Correct Estimate: true
Computed value: 0.963236   correct answer = 1.000000   Correct Estimate: true
Computed value: 0.854382   correct answer = 1.000000   Correct Estimate: true
Computed value: 0.115060   correct answer = 0.000000   Correct Estimate: true
Computed value: 0.528623   correct answer = 1.000000   Correct Estimate: true
Computed value: 0.996182   correct answer = 1.000000   Correct Estimate: true
Computed value: 0.000672   correct answer = 0.000000   Correct Estimate: true
Computed value: 0.168254   correct answer = 0.000000   Correct Estimate: true
Computed value: 0.525666   correct answer = 1.000000   Correct Estimate: true
Computed value: 0.004078   correct answer = 0.000000   Correct Estimate: true
Computed value: 0.000913   correct answer = 0.000000   Correct Estimate: true
Computed value: 0.000013   correct answer = 0.000000   Correct Estimate: true
Computed value: 0.549601   correct answer = 0.000000   Correct Estimate: true
Computed value: 0.000007   correct answer = 0.000000   Correct Estimate: true
Computed value: 0.926115   correct answer = 1.000000   Correct Estimate: true

Computed value: 0.292149   correct answer = 0.000000   Correct Estimate: true
Computed value: 0.000013   correct answer = 0.000000   Correct Estimate: true
Computed value: 0.000048   correct answer = 0.000000   Correct Estimate: true
Computed value: 0.882903   correct answer = 1.000000   Correct Estimate: true

The results shown in boldface show the 100 percent correct outcomes generated by the neural network. Not bad for a network constructed from scratch and without requiring the complex partial derivative computations associated with backtracking.

21.6 Summary

A relatively simple neural network with one hidden layer containing 25 nodes is constructed from scratch. It is trained with 150 pairs of diagnostic test results and associated labels with known correct results, over 1500 epochs. The network is then tested against 25 fresh test results, not in the original training set. The results are encouraging. The mean-squared error is shown to decrease to a small error as training progresses. All 25 test results produce the correct outcome.

Index

A

Abstract data types (ADTs)
    game, 123–127
    game, console implementation
        of, 128–135
    game of life, GUI implementation
        of, 135–138
    Go
        counter, 94–97
        counter package, creating, 98
        counter package,
          mechanics, 98–101
        implementing, 101–103
        OOP application, 109–121
        polymorphism, 106–109
        using composition, 103–106
        using classes, 91–93
    go.mod file, 138
    for grid, 128
    program output, 138, 139
    stacks, 141
Adelson Velsky and Landis (AVL)
      trees, 315
    avl package
        code implementation, 320
        deleteNode function, 334
        IntelliJ IDEA, 333
        main driver code, 332, 336
        map, 339
        rightRotate(node) function, 334
        rotateDelete function, 334
        Search method, 339

binary search tree, 315
    comparing set construction, 343
    concurrentAVLSet, 346
    dataSet slice, 343
    deletion, 318
    floatset package, 341
    insert and delete, 316
    insertion, 317
    interesting facts, 319
    tree Rotations, 316
Algorithm efficiency
    Big O, 55, 56
    searching array slices, 82–89
    slice of numbers, determining, 56–59
    sorting (*see* Sorting algorithms)
    speed efficiency, describing, 55
    using concurrency, 60–63
Artificial intelligence (AI), 549

B

Big O, 55, 56, 68
Binary searches, 87–89
Binary search tree (BST)
    deletion, 290
    generic implementation, 291
        data structures, 291
        delete, 294
        graphing, 297
        InOrderTraversal, 294
        insert, 294
        Main driver program, 310
        methods, 293

573

© Richard Wiener 2022
R. Wiener, *Generic Data Structures and Algorithms in Go*, https://doi.org/10.1007/978-1-4842-8191-8

W, X, Y, Z

Printed in the United States
by Baker & Taylor Publisher Services